PROTOPHYSICS OF TIME

BOSTON STUDIES IN THE PHILOSOPHY OF SCIENCE

EDITED BY ROBERT S. COHEN AND MARX W. WARTOFSKY

VOLUME 30

PETER JANICH

Philipps-Universität Marburg,
Institut für Philosophie, Marburg, F.R.G.

PROTOPHYSICS OF TIME

Constructive Foundation and History of Time Measurement

D. REIDEL PUBLISHING COMPANY

A MEMBER OF THE KLUWER ACADEMIC PUBLISHERS GROUP

DORDRECHT / BOSTON / LANCASTER

Library of Congress Cataloging in Publication Data CIP

DATA APPEAR ON SEPARATE CARD

LC no. 85-10768

ISBN-13: 978-94-010-8794-0 e-ISBN-13: 978-94-009-5189-1
DOI: 10.1007/978-94-009-5189-1

Published by D. Reidel Publishing Company,
P.O. Box 17, 3300 AA Dordrecht, Holland.

Sold and distributed in the U.S.A. and Canada
by Kluwer Academic Publishers,
190 Old Derby Street, Hingham, MA 02043, U.S.A.

In all other countries, sold and distributed
by Kluwer Academic Publishers Group,
P.O. Box 322, 3300 AH Dordrecht, Holland.

Translation by AAA Linguistics – Robert Brown

TABLE OF CONTENTS

EDITORIAL PREFACE

For protophysics, the fascinating and impressive constructive re-establishment of the foundations of science by Professor Paul Lorenzen, working with his colleagues and students of the Erlangen School, no task is more central than to.furmulate a theoretical understanding of the practical art of measurement of time. We are pleased, therefore, to have a new third edition of Peter Janich's masterful monograph on the protophysics of time, available in this English translation within the *Boston Studies.* We also look forward to the Boston University Symposium on protophysics in april of this year within which the full program of protophysics will be critically examined by German and American physicists and philosophers, supporters and critics.

We are also grateful to Paul Lorenzen for contributing his powerful instructive essay on the 'axiomatic and constructive method' which introduces this book.

March 1985

ROBERT S. COHEN
Center for the Philosophy and History of Science
Boston University

MARX W. WARTOFSKY
Department of Philosophy
Baruch College
City University of New York

PAUL LORENZEN

CONSTRUCTIVE AND AXIOMATIC METHOD

Mathematics is like a big building with many apartments. We have at least Arithmetic and Analysis, Algebra and Topology – and we have Geometry and Probability-Theory. Very often the tenants of these different apartments seem not to understand each other.

The Bourbaki movement promised a new unity of Mathematics by admitting only the axiomatic method of Hilbert as genuine mathematical.

This movement was quite successful – with one exception: the axiomatic foundations of Set Theory remained obscure.

But ignoring this difficulty which the mathematicians are accustomed to live with since about 1900, the axiomatic method leaves the mathematician without foundational problems. There are none in Arithmetic, because there are no non-Peano Arithmetics. There are none in Geometry, because the mathematicians believe that the physicists have the means of finding out whether the 'real space' is Euclidean or non-Euclidean. Finally there are no foundational problems in probability theory, because there are no non-Kolmogorow Probability-theories. In contrast to this happy state of "no problems by ignoring them", Constructivism tries to solve the foundational problems. It tries to justify, why mathematicians all over the world accept the Peano-axioms for Arithmetic and the Kolmogorow-axioms for Probability Theory. In the other cases Constructivism tries to find out, which axioms – if any – *should* we accept for Set-theory, for Geometry. The last question concerning Geometry would lead us away from Mathematics proper, so I shall go into this problem only later. But I would like to remark now that the whole foundational problem of Mathematics changes its outlook, if one begins to doubt the story that empirical physics has the means of deciding the case of Euclidean *vs.* Non-Euclidean Geometry. I would like to suggest, that you imagine a state of Mathematics without Geometry. We would be left with Numbers, Sets and Probabilities. We would have to develop Theories about these Entities without even being able to appeal to the paradigm of Geometry.

What would this look like?

Whereas the Axiomaticists, e.g., the Bourbakists, ignore all attempts to construct a foundation for the mathematical theories (or despise them even as 'pre-scientific') the Constructivists try to justify such axioms as those from Peano or Kolmogorow. The constructive method pleads for a cooperation of constructions and axiom-systems.

This cooperation is well-accepted in the case of *naive* Arithmetic and Analysis. 'Axiomatics' means here to define 'structures' as Groups, Lattices, Compact Spaces, Measure Fields.

The *structures* are defined by systems of sentence-forms, called *axioms*. The definition of such a structure is *justified* by showing that there are important *models*, which satisfy the axioms. The models, taken from naive Arithmetic or Analysis have here the priority. Only if they are important are the axioms accepted.

The language of the axioms may be restricted to pure logic, as e.g., in the case of group-theory or lattice-theory. But already a structure like 'archimedean ordered group' or the structures of 'topological spaces' use arithmetical or set-theoretical vocabulary.

This causes no difficulties because here the axiomatic method is applied *within* Mathematics. I shall call this *internal* axiomatics. The difficulties begin, if we now turn to the foundations. That is: we now no longer take for granted that such entities as natural or real numbers, and such entities as sets or functions are somehow 'given', are somehow already at our disposal. As if they were like flowers in a garden, which we have only to give names to – and then may begin to find out truths about them.

In Arithmetic the foundational problem is no serious problem. No mathematician seriously denies that it is easy to count, e.g., with such primitive numerals as I, II, III, . . . Everyone even understands the construction-rules for such numerals:

$$\begin{array}{l} \qquad \Rightarrow \mathrm{I} \\ n \ \Rightarrow n\mathrm{I} \end{array}$$

From such rules, to which may be added further rules for the construction of pairs, e.g.,

$$\begin{array}{l} \qquad \mathrm{I}, n\mathrm{I} \\ m, n \Rightarrow m\mathrm{I}, n\mathrm{I} \end{array}$$

(the constructible pairs are those with $m < n$) and for the construction of triples as for addition, an easy way leads to some first true assertions, e.g.,

$$\text{I} < n\text{I}$$
$$m < n \rightarrow m\text{I} < n\text{I}$$

This last sentence presupposes the logic of the conditional $\rightarrow$. No one seriously denies $\neg\, n < \text{I}$, if he accepts negation $\neg$ at all. Look, this is not a matter of choosing 'axioms'. If someone would propose, e.g., $\neg\, n < \text{II}$ instead, he would be refuted that very moment because $\text{I} < \text{II}$ is true.

Also, e.g., $m < n \rightarrow m\text{I} < n$ would be ridiculous. Because $\text{I} < \text{II}$ is true, but $\text{II} < \text{II}$ is not.

For such arguments we do not need *formal* logic; but we have to know how to argue with meaningful sentences (here of the form $m < n$) which are composed by logical particles such as $\rightarrow$ and $\neg$. 'How to argue' means the dialogical rules for the logical particles.

Those rules, e.g.,

$$\begin{array}{r|l} & A \rightarrow B \\ A? & B \end{array}$$

for the conditional $\rightarrow$ have been developed within contructive logic.

The controversy between the classical and intuitionist logicians, e.g., whether $(\neg A \rightarrow \neg B) \rightarrow (B \rightarrow A)$ is logically true (or only the converse), turns out to be an empty struggle. You have to justify in any case the choice of a general rule for dialogues. For intuitionist and for classical logic this can be done by proving a corresponding Gentzen cut-theorem. Classical logic turns out to be a convenient simplification of intuistionist logic – in the case of junctors only you have just to omit all adjunctions $\vee$ and conditionals $\rightarrow$ (and to redefine $A \vee B$ by $\neg(\neg A \wedge \neg B)$ and $\text{A} \rightarrow \text{B}$ by $\neg \text{A} \vee \text{B}$).

The Peano-axioms turn out to be true sentences in constructive Arithmetic. The ω-incompleteness, proved by Gödel, shows that not all constructively true sentences are logically derivable from the axioms. This is little wonder. A universal sentence $\bigwedge_x A(x)$ is constructively true, if $A(n)$ is true for all n. But in order to derive $\bigwedge_x A(x)$ *logically* you have to derive first $A(x)$ with a free variable. So one should have expected the ω-incompleteness. But the Peano-Arithmetic is ω-complete, if you exclude multiplication. The point of Gödel's proof was to show that the Peano-Arithmetic with addition and multiplication only – without higher forms of inductive definitions – already exemplifies the ω-completeness, which was to be expected in general.

The quarrel between the axiomatic and the constructive method in Arithmetic is merely a verbal one. As the Peano-axioms are constructively true there is no point in insisting that mathematics does not begin before you have put down some axioms.

The quarrel becomes serious if we go to Analysis. From Cauchy to Weierstrass the mathematicians of the last century have succeeded in defining the real numbers and in proving their fundamental properties, especially the completeness-property. Definitions and proofs are not arithmetical, because they make use of 'sets' and certain theorems about them. Sets were *not* defined and their fundamental theorems, e.g., the comprehension – principle

$$\bigvee_{\mathscr{S}} \bigwedge_x (x \in \mathscr{S} \leftrightarrow \mathscr{C}(x))$$

for all sentence-forms $\mathscr{C}(x)$ were *not* proven.

But it was already clear from Frege that sets should be defined in the following manner: two sentence-forms $A(x)$ and $B(x)$ for which

$$\bigwedge_x (A(x) \leftrightarrow B(x))$$

holds, represent the *same* 'set'.

If we use $\{z \mid A(z)\}$ resp. $\{z \mid B(z)\}$ for the 'sets' represented, we have as *definition*

$$\{z \mid A(z)\} = \{z \mid B(z)\} \leftrightharpoons \bigwedge_x \cdot A(x) \leftrightarrow B(x).$$

This is a definition by abstraction. After the abstraction, we may define ϵ by *e*.

$$x \in \{z \mid \mathscr{C}(z)\} \leftrightharpoons \mathscr{C}(x)$$

From this the comprehension-priciple immediately follows. But with one remarkable restriction. The formulas A, B, C here should *not* contain any quantifier ranging over *sets*. Otherwise we would have no definition of sets.

So the constructive approach does not yield the truth of the unrestricted comprehension principle, that is the comprehension principle without the restriction of $\mathscr{C}(x)$ to formulas without quantifiers ranging over sets. Con-tructively we get only so-called 'predicative' comprehension. But modern axiom-systems for Set-theory contain always axioms which imply at least some unrestricted cases of the comprehension-principle. Naive and axiomatic Set-theory are 'impredicative'. So we do not have a straight forward proof that the axioms of Set theory are constructively true. That is the crucial difference with Arithmetic.

The Hilbert attempt to add to the axiom-system a logical calculus and then to prove that the resulting *formal system* is at least formally-consistent (i.e., for no formula A is A and $\neg A$ derivable in the formal system) is still with us. But it seems – unfortunately – very difficult.

This difficulty does not mean, however, that a responsible mathematician could not do Analysis as long as the consistency-problem is pending. Most mathematicians just don't care (perhaps most even don't know), that their formal system *may be* inconsistent. If it were inconsistent, it would be useless to spend one's time trying to find derivations – all formulas would be derivable.

This danger of irresponsibly wasting one's time can be avoided by doing Analysis constructively – instead of formalizing the Cantorian muddle of 'naive' set theory.

Constructive Analysis means to begin with constructive Arithmetic, to proceed to the rational numbers – and then to *construct* sentence forms $A\,(r), B\,(r), \ldots$ with a free variable ranging over rationals. The sentence-forms are built up from Arithmetic only, i.e., from inductive definitions, and combined with logical particles, the quantifiers ranging over the natural or rational numbers only. Restricted (predicative) comprehension alone leads then to the definitions of 'sets' of rationals, and of 'real' numbers. Real numbers are again defined by abstraction. The equality $\mathit{fin}\ r\,A\,(r) = \mathit{fin}\ r\,B\,(r)$ is defined by $\{r\,|A(r)\}$ and $\{r\,|B(r)\}$ having the same *lower* class of rationals. The restriction of having no set-quantifiers in the sentence-forms implies to have also no real-number-quantifiers in the comprehension-principle.

The point of Constructive Analysis, as is shown in my book ***Differential und Integral. Eine konstruktive Einführung in die klassische Analysis*** (Frankfurt 1965), is that in spite of this restriction Analysis may essentially proceed as usual. For all applications of Analysis – outside of modern pure Mathematics – the constructive methods of Analysis are sufficient.

The further replacements of classical logic by intuitionist logic or of the functions by recursive functions, require much greater changes in Analysis. Therefore 'constructive' Analysis is often said clumsy and inconvenient. But this is at most true for these further restrictions. 'Constructive' Analysis in my sense, which merely restricts the comprehension-principle has the same 'elegance' as Analysis on an axiomatic set-theoretical basis. It just avoids the danger of inconsistency from the beginning. So that no consistency proof is required either. The advantage of constructive Analysis in all its forms is to get rid of the pluralism of axiomatic systems. Even if we would have a consistency proof of ZF (Zermelo-Fraenkel) we would have – on the basis of the work of Gödel and Cohen – a consistency proof of ZF $\wedge$ CH and of ZF $\wedge \neg$ CH (with CH for the Continuum Hypothesis). And there would be no rational decision between these two formal systems. We still would have the present situation of a plurality of irrational choices – where only indifference, called 'tolerance', gives the illusion of serving a common purpose. The variety of 'constructive' investigation is not pluralism, but indicates merely different emphasis. Constructivism in the non-recursive sense of my book embraces recursive Constructivism, because the restriction to recursive functions merely singles out these functions as especially important. Intuitionist logic on the other hand embraces classical logic, as I have observed already. So the restriction to classical logic is not an irrational choice, but concentrates – for the sake of simplicity – on classical existence $\neg \bigwedge_x \neg A(x)$ instead of $\bigvee_x A(x)$. Intuitionist non-recursive Constructivism would be the all embracing theory – the other 'Constructivisms' are subtheories.

In contrast to this peaceful state of affairs, we have a pluralism of axiomatic theories, which contradict each other as CH and $\neg$ CH. Mathematicians are used to such contradictory pluralism since the time of Non-Euclidean Geometry. I shall come to the geometrical problem later. I continue with an attempt at understanding the remarkable fact, that in our pluralistic world there is only *one* axiomatic theory of probability, that there is no Non-Kolmogorow Probability-theory. It seems so simple to define 'probability', because within Analysis – now it does not matter whether constructively or axiomatically done – we may define, say, Kolmogorow-fields as σ-fields of sets with a σ-additive normed measure. That is not the problem. The problem is why are the Kolmogorow-fields called 'probability-fields' – and how is the term 'probability' defined, so that we can *justify* its *application* in all the areas covered by modern statistics. This problem of justifying the applicability is at stake, when one speaks of the justification of the Kolmogorow-axioms.

Let me start with descriptive statistics, where a set (population) of N elements $c_1, \ldots, c_N$ is given. Let it be known that n-elements of the population satisfy a certain sentence-form $A(x)$. The frequency of A – mostly called 'relative frequency' – is defined as

$$S(A) = \frac{n}{N}.$$

Elementary Arithmetic yields

I $S(A) = 1$, if all elements satisfy A
II $S(A \vee B) = S(A) + S(B)$, if A and B are disjunct.

Now an element shall be chosen. We don't know, which. But let us call it already 'c'. Then we have predictions about $A(c)$. Predictions are done with modalities. If $S(A) = 1$, then $A(c)$ will be *necessary*. If $S(A) = 0$, then $A(c)$ be *impossible*. For $0 < S(A) < 1$ we get that $A(c)$ is *contingent* (i.e., neither necessary nor impossible).

Now it is suggested to use comparative language – and to say that $A(c)$ is the more necessary the more $S(A)$ equals 1. Or then, quantitative language – and to say that $A(c)$ has the 'degree of necessity' $S(A)$, and instead of 'degree of necessity' you may say 'probability'. But this does not make sense in general. Take the population of a town and let the *frequency* of females and males be 1/2. Now choose the first person you see in the morning. The *probability* that it is female should be 1/2. Well, at least in the case of a husband this is absurd, as he sees his wife first each day, by necessity, so to say. This example shows that the element c has to be chosen *by chance*. That means in order to justify the talk of probabilities we first have to define first 'chance'. This can be done with the help of game devices, such as dice or roulette. We will call such devices *chance-generators*.

In the *discrete* case we require that a *change-generator* is a device with

(1) *Uniqueness*: Each use of the device yield as *result* exactly one of finitely many sentence-forms $E_1, \ldots, E_m$ ('elementary events')

(2) *Repeatability*: After each use the device is in the same state as before the use

(3) *Indistinguishability*: According to no causal knowledge is there a reason to distinguish between the results before use.

This definition of chance-generators does not belong to mathematics. It is a prescription for the makers of such devices. For the following I assume that engineers do understand this language – and that our industry is able to realize such chance-generators. No real chance-generator will be perfect – but they are good enough to justify the notion of chance-generator as defined by its *uniqueness, repeatability* and *indistinguishability*.

The results of a chance-generator are contingent. We now propose to define a *probability* p for the sentence-forms $E_{x_1} \vee \ldots \vee E_{x_r}$ (i.e., the adjunction of the elementary 'events' such that

I $\quad p(A) = 1, \quad$ if A necessary
II $\quad p(A \vee B) = p(A) + p(B) \quad$ if A, B are incompossible
III $\quad p(E_1) = p(E_2) = \ldots = p(E_m)$

We require that p behaves as a frequency (I–II) and add III because of the indistinguishability. Of course, I–III yields immediately:

$$p(E_{x_1} \vee \ldots \vee E_{x_r}) = \frac{r}{m}$$

That is, we have a Laplace-field of probabilities. The definition of chance-generators surely justifies III. But why do we require that the 'degree of necessity' is a frequency? The reason lies in Bernoulli's law of large numbers. If we use the chance-generator in a series of length L, we have on the one hand $p(A)$ according to I–III, but we may observe the frequency $\rho_L(A)$ on the other hand. There are m^L results of such a series. Because of the repeatability they are indistinguishable. So we get a probability p_L for all adjunctions of them, especially for the sentence-forms $|\rho_L(A) - p(A)| < \epsilon$ for each positive ϵ. Bernoulli's law says

$$\lim_{L \to \infty} p_L(|\rho_L(A) - p(A)| < \epsilon) = 1,$$

Which is read – very roughly – as saying that the probability $p(A)$ *is* the frequency $\rho_L(A)$ 'in the long run'. The purely mathematical proof of Bernoulli's law justifies defining 'probability' with the help of I–II (which are true for frequencies). I–III are a definition, *not* an 'axiom-system'.

In the continuous care of chance-generators there are no elementary events. A roulette wheel is arbitrarily divided in a finite number of intervals of *equal* length. If the total circumference has the length 1, the probability of any interval is just its length. For a 2-dimensional device we would have correspondingly the area of rectangular intervals as probability. As it is arbitrary to be restricted to such intervals the probability is by definition extended to the Lebesgue measurable subsets.

This means that II is strengthened to σ-additivity for sequences A_*

II $\quad \sigma p(\bigvee_r A_r = \sum_r p(A_r), \quad$ if all A_{r1}, A_{r2} are incompossible.

Continuous chance-generators lead thereby to Lebesgue-fields of probabilities in the n-dimensional number-space.

Once we have technically realized chance-generators with their (Laplacean or Lebesguean) probability-fields we easily get further probability-fields by combining the generators to *aggregates*. If the generators are causally independent (that is again a matter of engineering) we may define the product-field by using the generators simultaneously. This product is again a model of the *Kolmogorow-axioms*. I and II σ. Once we have chance-aggregates other operations lead to further probability-fields, namely

(1) the *relativation* (which means to define $p_B(A) = \dfrac{p(A \wedge B)}{p(B)}$ for any B with $p(B) > 0$) and

(2) the *engrossment* by which the σ-field is homomorphically mapped into a new σ-field.

Continuous examples are given by all sorts of false dice. I'll take a quadratic prism

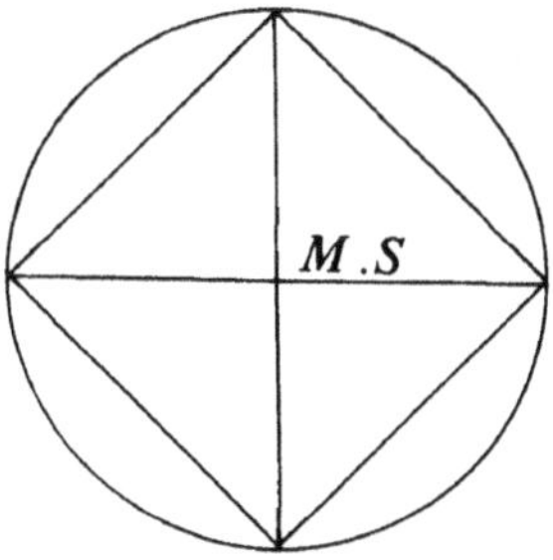

with a barycenter S outside the geometric center M. If the probability-field for falling on an interval of the circumference (circumscribed only in thought) is a Lebesgue-field, the probability field for the 4 sides will be a discrete *engrossment*, but not a Laplace-field. Obviously in this way we get *all* discrete fields. Finally Kolmogorow has shown that the course of events in stochastic processes (which are at each moment only determined by a probability-field) are themselves again the 'events' of a probability-field.

Therefore since Kolmogorow 1933 the 'mathematical probability-theories' is nothing else than the theory of σ-fields of sets with a σ-additive normed measure. May I suggest to be somewhat more careful? I propose to call an arbitrary model of the Kolmogorow-axioms a 'Kolmogorow-field'. All the models which we get by starting with Laplacean and Lebesguean fields and then applying all the operations due to relativation, engrossment, products and stochastic processes, satisfy additional 'axioms'. In all applications we have it to do with the Borel-field of topological spaces. It has been shown that the class of so-called 'polish' spaces (which are separable and completely metricizable) is *sufficiently large* to be closed with respect to the above operations on probability fields.

The class of polish Kolmogorow-fields is on the other hand – in contrast to the class of all Kolmogorow fields – provably *not too large*: all these fields are representable as limits of *discrete* probability fields. The notion of limit here has been made precise by H. Cartan, but has been named by him 'vague convergence'. This approximation-theorem justifies calling all models of normed σ-additive measures of the Borel field of polish spaces simply 'probability fields'. We are back to *internal* axiomatics, where the models come first. The relevance of the approximation theorem comes, of course, from the applications.

In all applications of probability theory to natural processes ('where the causal explanation does not succeed') we try the fiction that the process runs *as if* determined by a hidden and unknown chance-aggregate. We do not try to find the elements of this fictitious aggregate – we try only to find useful approximations of the probability-field. That is the task of statistical test-theories. With this example of a constructive foundation of the axiomatic probability-theory (namely by constructing chance-aggregates as models) I would like to conclude the subject of mathematics proper. The constructive method opens the way to overcome the irrational pluralism of merely axiomatic Mathematics.

Yet probability theory can be regarded as not belonging to mathematics proper. The objects of this theory are fields of 'events'. These 'events' – according to the reconstruction given above – are only adjunctions of forms of sentences, of sentences however about chance-generators. And chance-generators are technical machines. Probability theory therefore makes sense only because we produce and use those machines.

To be sure also numbers (e.g., /, //, ///, ...) make sense only because there are objects which we count by numbers, but these objects do not belong to Arithmetic itself.

Probability theory therefore may be regarded as a first step from (pure) mathematics towards physics. In contrast to Arithmetic which only deals with symbols, Statistics deals with 'reality'. The access to a theory about reality is provided by a prescientific technique of the production and use of chance-generators.

Following this pattern the access to geometry also has to be given via a prescientific technique. This apparently marks first steps into physics. We therefore call this field 'protophysics'. In terms of tradition we have to deal with the theories of space, time and matter. Especially the fundamental magnitudes of empirical physics: length, duration and (inert) mass have to be defined. The theory of space is the oldest of these theories and grew out of – as its name 'geometry' says – a technology on earth.

Any production of tools or machines concerns 'matter', especially pieces of wood, stone or metal. For the sake of illustration and in order to recall the sense of the greek ὕλη for matter, we choose pieces of wood. We dispose of the technique to make a part of the surface of a wooden piece flat. We cannot produce perfect planes but sufficiently good approximations. As is well known any two of three flat surfaces fit each other and are freely moveable when brought into contant. Additionally, wedges (i.e., bodies with two flat surfaces intersecting each other) can be produced; the two flat surfaces of which form a right angle. We write $E_1 \perp E_2$. If we produce three plane parts of a surface pairwise orthogonal and put this body on top of the flat surface E of another body we obtain an edge g orthogonal to that plane. We write $g \perp E$. By repeating this procedure we obtain another edge $g' \perp E$, which may be called 'parallel to g', in short $g \parallel g'$. By production of flat surfaces and rectangular edges, ideal norms of homogeneity and symmetry are 'realized'.

The problem of a constructive foundation of geometry then consists in the explication of norms in such a way that a *complete* system of geometric axioms follows from them.

This problem has been solved in the meantime by the books of Rüdiger Inhetveen (*Konstruktive Geometrie. Eine formentheoretische Begründung der euklidischen Geometrie*, Mannheim/Wien/Zürich 1983) and of mine (*Elementargeometrie. Das Fundament der analytischen Geometrie*, Mannheim/Wien/Zürich 1984). The second part of protophysics, the theory of time-measurement, has been worked out in this book by Peter Janich.

PREFACE TO THE SECOND EDITION

The first edition of this book was intended to reach a very special objective. Based on Hugo Dingler's philosophy whose statements concerning the foundations of physics were interpreted and defined by P. Lorenzen in two articles, it was my intention to actually execute one part of the protophysical program; namely to explicitly formulate as a first part of the reconstruction of physical terminology those established postulates (*Festsetzungen*) which in linguistic and non-linguistic respects are at the basis of the art of measurement as practiced by physicists. As the guiding principle of these reconstruction efforts I meant to apply the principle of methodical order according to which – on the basis of extrascientific speech and action, i.e., independently of the results of modern physics – all linguistic and non-linguistic steps required for an operative definition were to be enumerated in such a way that in doing so neither gaps nor circles would occur. For this purpose, the measurement of time presented itself as the simplest case.

This special objective led to certain shortcomings in the first edition which I hope to have now remedied in this second edition. In the first edition, no effort was made to refer to current scientific theoretical discussions. Although, at that time, potentially competing theories had been suggested by H. Reichenbach, C. G. Hempel, R. Carnap and others and although the protophysical tried to avoid their specific shortcomings, this counterposition to the theories of concepts which originated from the tradition of the Vienna Circle was in itself not the topic of this investigation. Thus, the critics of this book did not pay attention to its claim that here a better alternative than the accounts mentioned above is being offered. I myself am largely responsible for this misinterpretation of the protophysical account, and therefore I added a new first chapter to the second edition in which I try to refer to counterproposals and want to explain and prove the claim that I indeed offer an improvement upon competing theories which are presently under discussion.

Unfortunately, the renunciation of any attempt to establish an explicit reference between protophysical chronometry and the most important alternative theories led to another consequence – it was seen as an uninterrupted continuation of Hugo Dingler's philosophy. Notwithstanding the question whether Dingler's approach itself might already have become a mere caricature in this discussion, the identification between protophysics and Dingler's approach was misleading in its most essential point. To be sure, in the first edition I had already stressed that the results (about which, by the way, no doubts are being raised) of modern empirical physics could not be used as objections against protophysics. But this methodological state of affairs has nothing in common with the claim for absoluteness of Dingler's proofs. But there is not only a difference in principle between Dingler's claims and my own. The execution as well is totally different – Dingler, in my mind, has not given an operational proof of time measurement at all. Therefore, in an introduction to the otherwise largely unchanged section about Dingler, I now have tried to point to the differences between the protophysical approach and Dingler's approach. Although I continue to present Dingler's remarkable methodological suggestions, this will no longer allow anyone to criticize Dingler and to actually mean protophysics or vice-versa.

As further alterations from the first edition, I would like to mention additions in the discussion of Lorenzen's writings and, most importantly, the completely new methodological arguments presented in Chapter II. Here, I owe many suggestions to conversations with F. Kambartel, and also to criticism by G. Böhme and A. Kamlah and others. I hope that the claims of protophysics which were so frequently misunderstood will now be seen and judged in a better light. As the criticism of protophysics and the responses they elicited have now been published in a separate book the anticipation of objections by those readers who are not familiar with this discussion can be limited to a few major arguments.

The construction of a chronometry in Chapter III is now preceded by a largely new section where the expectations of a theory of time measurement are named – expectations which can be proven in a non-circular manner and which must be formulated beforehand exactly in those cases for which a theory of time measurement is not meant to be an analytical distillation from accepted physical theories where just the unreflected preconditions of a practice of time measurement have already settled in. The chapter on clockfree kinematics remained unchanged except for a few small technical corrections. But in a new section I have discussed in detail the choice of repeatable events which, so to speak, form the raw material for any art of time measurement and its theory. And in a clearer manner than in the first edition, the quality of relative repeatability of events as a criterion for class formation is integrated into the proof of unambiguousness of the chronometric principle of homogeneity for uniform motion. But now this proof is a direct one and no longer established indirectly as in the first edition. The subsequent example of the elimination of irregularities in clocks was written anew because the example in the first edition obviously gave rise to misunderstandings and had suggested a confusion between the process of realisation and exhaustion. The historical fourth chapter finally has been expanded and now includes a discussion of the Augustinian theory of time as well.

I hope that the improvements mentioned will also testify to the injustice of suspecting protophysics of being anachronistic, as is mainly suggested by the rejection of protophysics by those authors who, in their pursuit of the ideals of completeness, have devoted two or three critical sentences to protophysics. Apart from the fact that a reproach of anachronism must argue by using the results of post-classical physics and that it thereby is circular with respect to a claim of proof for physics, I in fact never concerned myself with a restoration of classical physics. Indeed, in all three areas of length-, time-, and mass-measurement, classical physics has just as many gaps in its proofs and definitions as post-classical physics, and that in its most recent formulations as well as in Newton's. That the origin of protophysics nevertheless follows the classical conceptual triplet of length, time and mass has neither historical nor metaphysical reasons but rests solely on the fact that an experimental physics in the systematic construction of its instruments must first begin with the spatial forms of bodies, and then deal with the temporal forms of processes; in short, this sequence is a consequence of the principle of methodical order.

In respect to the discussion which developed since the appearance of the first edition between the theorists of science, and in respect to the controversy between the followers of Popper and Kuhn, or in respect to the actuality of the Sneed-Stegmüller-discussion, a new kind of suspicion of anachronism might be likely to arise; namely, that protophysics might be outdated although not in respect to the history of physics but in respect to the history of the theory of science. To this one could object that achievements of the more recent discussion must seem to be revolutionary only to those who come from the philosophy of the Vienna Circle. Insights like those that physics does not only consist of its theories but also of a practice of persons who are acting within the historical framework and whose acts are intentional, was from the outset a constituent of protophysics. Also, a methodology is not proven anachronistic by the mere fact that it collides with fashionable radical postulates of liberalism. Reforms as well as revolutions always require a certain amount of continuity in human intentions. A methodology with the principle of methodical order which resembles a rigid claim for exactness does not become superfluous by the fact that representatives of anti- and meta-methodologies exclude the problem that every bit of knowledge of nature discovered through experiments, i.e., through instruments, is created by the success or failure of human intentions which were realized by the instruments. Our knowledge of nature is therefore always essentially a knowledge as to which of our intentions have become successful or remained unsuccessful in the artificial creation (*Herbeiführung*) of processes and states, or in the explanation of something that was caused in an unplanned manner, and it therefore presupposes an explicit knowledge of these our intentions.

PREFACE TO THE THIRD EDITION

Protophysics has become a well-known technical term of the philosophy of science in German speaking countries. None of the newer German handbooks or dictionaries on the subject leaves the catch-word 'protophysics' out. It stands for a constructive (or, as its critics use to say, 'constructivist') program for the foundation of physics. The development of the program itself and its pursuance has led to an operational foundation of geometry, of chronometry and of major parts of dynamics and of a theory of waves. Additionally the methodology of physics together with related parts of action-theory has been developed, and a rather broad discussion took place about the relation between contructive and analytical philosophy of science, and between normative protophysics and empirical, mainly relativistic physics as well.

Protophysics counts as a main topic of the so-called Erlanger Schule of constructivism and therefore marks a position in the field of philosophy of science as it is discussed mainly by philosophers. However theoretical physicists also show some interest which led to a few publications in German language although this discussion stands more at the beginning. Merits have to be attributed to many critics of the program as well as of its fulfillment in detail. Their questions and objections concerning the philosophical background, the constructive methodology and the particular theories of measurement of length, time and mass have been a fruitful challenge and led to a few smaller conferences and to a series of publications.

On the other hand barriers of language restricted the discussion of and the interest in protophysics to the German speaking area. The only English publications consist of a short characterisation of the program by P. Lorenzen (Normative Logic and Ethics, 1969), of a paper about mass-measurement by the author (Janich, 1983) and a critique of the protophysics of time by J. Pfarr which was presented at the Boston Colloquium for the Philosophy of Science 1978 and was published, together with a response of mine, in the Boston Studies in the Philosophy of Science Series, Vol. 82. The third edition of my book on protophysics of tiem now in English is supposed to overcome the language barrier. The work on the translation and the edition has, for a variety of reasons, lasted since almost ten years. This long time brought however a systematic advantage: whereas the second edition of this book on time-measurement still had to rely on a hypothesis and a promise concerning the presupposed protophysical foundation of length-measurement, two books upon constructive geometry have been published in the meantime (R. Inhetveen, 1983; P. Lorenzen, 1984). To be sure, a few minor differences regarding some points of detail between these authors and me still have to be settled but these do not concern the program in general or the opinion that a constructive foundation of geometry is now available and provides a basis for the protophysics of time.

Finally the book may raise interest in philosophical questions which are to be discussed at a Boston Colloquium for the Philosophy of Science in April 1985 and which shall cover the entire domain from geometry through mechanics to special and general relativity. It may additionally draw attention of the English speaking public to a piece of present German philosophy which will be presented by another book to be published at the same time (Constructivism and Science. Essays in Recent German Philosophy, ed. by Robert S. Butts, London/Ontario, Canada).

Looking back at the delayed history of the English third edition of this book I would like to thank both the editor and the publisher and all helpfully involved persons for their patience and support.

November 1984 PETER JANICH

CHAPTER I

ON THE PROBLEM OF CHRONOMETRY IN THE PRESENT-DAY THEORY OF SCIENCE

1. INTRODUCTION: ESTABLISHMENT OF A REFERENCE TO KNOWN POSITIONS IN THE THEORY OF SCIENCE

The present volume develops a theory of chronometry for physics; in doing so it pursues the end of itself providing a small and fundamental piece of the theories of the special brench of science, physics. Independently of whether or not this end can be realized successfully, professional physicists are intended as the primary addressees of the protophysics of time. The definition of 'clock' or the provision of a metric time concept thus has its place in the introduction or in the first chapter of a textbook of mechanics. For this, the relationship of an operative foundation of chronometry in a protophysical theory to technical physics or to theories of technical physics must be clarified. This takes place in Chapter II, in the context of a discussion of methods.

Notwithstanding this correspondence of protophysics with a physics concerned about its foundations, it is, however, an historical fact that, at least institutionally, a division of labor between professional physicists and theoreticians of science has prevailed. Overlooking a few exceptions, physics has delegated concern about its foundations: Still without criticism and evaluation, this can be understood initially as an event analogous to the division of labor between theoretical, experimental and applied physicists. In fact, however, this delegation also leads to the fact that instead of appealing to the intended addressees from technical physics, a proposal such as the one made in this book must in the first place actually address itself to theoreticians of science, because it will be read and discussed by them above all and not by professional physicists.

Now it is a triviality of the history of the theory of science that the theory of science or philosophy of physics has become autonomous, at least insofar as analyses of physics begun in the second half of the previous century have led to an abundance of questions of logic, of theory of definition and theory of truth; these questions can be discussed in themselves and with an increasingly diminishing visible relation to professional physics. This detachment of the theory of science from professional physics – a detachment widely recognized today and one that also could not be undone due to more recent endeavors of the sciences of science (sociology of science, history of science, history of institutions, psychology of researchers, etc.) – entails that in addition we must first establish the relation of a systematic proposal for defining a physical basic concept addressed to physicists to the main themes in the theory of science. The first chapter is dedicated to this task.

Admittedly in asserting that we must additionally establish a reference of the present protophysical account to known positions in the theory of science, we are neither asserting that these positions themselves will or can claim for themselves no relation to professional physics nor are we stating thereby that the present protophysical account has been developed with a view to physics alone, ignoring known results in the theory of science. Establishing a relationship of protophysics to the theory of science of other provenances is thus not merely a didactic effort to facilitate the reader's incorporation of protophysics into the series of other accounts: Rather it is above all a matter of making clear that the protophysical attempt has arisen against the background of alternative conceptions and that it claims to avoid their disadvantages. It will be seen already in the following pages that the protophysical account, which leads to a *normative theory of measurement*, cannot get along without the results of a theory of science that proceeds descriptively. Thus in this chapter we will illustrate to what extent a theory of science that merely analyzes and describes physics, when it appears with the claim to be the whole theory of science, is abbreviated and, when it does not appear with this claim, is to be considered as a partial endeavor that must be extended by a protophysical part in order to make possible, say, a founding of chronometry for technical physics.

In detail, this discussion will begin with a presentation and clarification of an affirmative basic attitude of the so-called 'analytic' theory of science, the characteristics of which show in several approaches to physical concept formation; it will lead finally to a criticism of theories of the formation of metric concepts in general and of the concept of time to particular.

2. AFFIRMATIVE THEORY OF SCIENCE AND THE LANGUAGE OF PHYSICS

The history of the modern philosophy of the exact sciences begins almost exclusively as a theory of scientific language. It can remain open here, to what extent the attempted conquest of traditional philosophy in the from of linguistic criticism, has played a role in this, say, as a consequence of

Wittgenstein and of analytic philosophy. In the history of the exact sciences themselves, the questions of a theory of scientific language that are still central today are delineated.

Already Euclid, following upon Aristotle's theory of science (however, not a theory strictly adhered to by him), left western science the model of a scientific theory in the form of his axiomatic geometry, a model still valued today. But thereby he also essentially laid the basis for the problems which accompany mathematics and physics up to the present. By this is meant above all the question of what status the definitions preceding Euclid's postulates and axioms have. Already ancient commentators objected to their deficiencies and sought for alternatives. These definitions did not satisfy the requirement of fixing defined words by reducing them to non-geometrically defined or definable words; they do not follow the sequence of a systematic construction of a terminology, in which in later definitions, as far as possible, only the previously available words occur; finally, they play as good as no role in the accompanying geometry: They are not employed as arguments either in theorems or in the proofs of theorems. Thus the dispute over the status of the parallel postulate which essentially determines the further history of geometry, could not be carried out upon the foundation of an explicitly defined geometric terminology that left no ambiguities open. The consequence was that, to a large extent, syntactic considerations supplanted semantic ones, until finally David Hilbert with his proposal of a formalistic understanding of geometry, completely excluded the question of the meaning of geometric basic terms from mathematics in favor of a syntactic criterion. This confusion of the sense and meaning of scientific statements,[1] which was first established by the deficiencies of Euclid's geometry, only apparently obviated in the analytic reinterpretation of geometry by Descartes in the domain of algebra and elevated to a program in Hilbert's 'Axiomatisches Denken', has endured until the present day; it has also allowed the theory of science to become primarily a theory of scientific language.

The same deficiencies which show themselves in the geometry of Euclid and repeated in a similarly successful axiomatic theory, namely the mechanics of Isaac Newton. There also the definitions of absolute space and absolute time as well as of mass, which definitions are premised before the axioms, are insufficient and, for the theory following upon them, they are also partially irrelevant. (Cf. for example Nagel, 1961, p. 203 f.) In strange parallel to the history of Euclid's geometry there developed a technically highly successful physics, which (analogously to the apparent disappearance of definition problems through the translation of synthetic geometry into algebra) believes itself to have, if not solved, than at least to have evaded its foundational problems by means of analytical mechanics (Euler, d'Alembert, Lagrange, Hamilton and Jakobi).

The series of such significant developments in the exact sciences, in which old foundational problems are not solved but rather expelled from the domain of a theory or a discipline, could be continued with the subsumption of mechanics under electrodynamics, the reinterpretations, above all, of the status of the (deficiently defined) basic concepts of classical physics through relativistic physics, up to the wave mechanics of E. Schrödinger. In a rough overview of the history of physics and its basic concepts all developmental steps which pass for radical are attended with an abundance of problems of the linguistic and conceptual representation of their results. Thus, in addition to other possible reasons, it is the history of the exact sciences themselves that predestines the theory of science *prima facie* to be a theory of scientific language.

The first results laboriously obtained above all by the members and followers of the Vienna Circle, and among these particularly by Rudolf Carnap and C. G. Hempel, (above all: Carnap, 1926 and Hempel, 1952, Carnap, 1966) also lie in this area then. We can state in advance that these endeavors have made possible an exactness of speech about natural science which one cannot lose. Nevertheless one can also ascertain that the theory of science that pursues the program of logically analyzing the language of physics has gone in a direction which itself produces new problems without already having solved the old ones. To be able to go into these problems in more detail, we must point to an aspect of history that the linguistic-analytic program has experienced.

Among the self-evident presuppositions of this history is the fact that theories in mathematics and physics had already been elaborated. The open problems of definition in set theory, geometry, mechanics and electrodynamics belonged so to say to the basic knowledge of this historical starting point. If one takes into account the fact that there were influences from the criticism of traditional philosophy, in this case more precisely of Kant's doctrine of the synthetic *a priori*,[2] (a criticism according to which the doctrine that there can only be analytic and empirical statements . . . belonged more to the presuppositions than to the results of the analysis of the language of physics) then the problem of how to obtain the existing propositions of physics had to be solved solely with the help of experience and logic or mathematics. Whether now in the radical form of the theory of observation sentences (*Protokollsätze*) (Cf. Neurath, 1932; Carnap, 1952) or of a conception which began with propositions about individual experiments and was *plagued* less with terminological scruples (Cf. Bridgman, 1952) attempts to solve the definition problems of a physical theoretical language on the basis of an observation language foundered. With some authors this foundering encouraged the view that there was no 'pure observation language'; instead even the simplest observations would be describable only in theory-laden propositions. Here it is obviously a question of in what sense the 'existence' of an observation language is disputed. Certainly the fact that the everyday language of the physicist is saturated with scientific terminology is no proof. For an explicitly defined observation language building upon the exemplary definition of predicates, independence from any physical theory can be secured. This assertion can also be maintained against the view of Kuhn and Hanson. From the apposite remark that, with a 'pardigm shift' in the natural sciences, (possibly all relevant) expressions change their meaning, it certainly does not follow that the complete construction of a technical terminology upon an ordinary language, itself partially reconstructed to this end (which, in fact, scarcely occurs in the history of science), would be theory-neutrally impossible. (Cf. Stegmüller, 1973, p. 165). Expressions of this are the well-known development of the Carnapian position intended is the gradual liberalization of the program of reducing definitionally the language of physical theories to observation language, which was then withdrawn to requiring the translatability of theoretical languages into an observation language and finally even had to allow uninterpretable terms. (Cf. Stegmüller, 1970) which proceeded in three phases – or the objections of Reichenbach (cf. Reichenbach, 1928, § 4) and then of Hempel (Hempel, 1952, p. 31) to empiricist (in the narrower sense) variants of the conception of theoretical languages, and the formal working out of the connection of linguisticanalytic accounts and Kuhnian criticism in the theory of Sneed. (Sneed, 1971).

Without going into the details (available in textbooks in the theory of science) we can designate as the interesting characteristic of this development here the fact that finally, in the face of all further differences, it was agreed *to no longer discuss problems of the meaning of physical terms independently of problems of the validity of the propositions formulated with them*. Thus C. G. Hempel, who repeatedly stresses the coincidence of concept definition and formation of theories, says that it is "advisable . . . to give up the view that the propositions of a theory divide into two classes distinguished by their epistemological features: propositions secured by linguistic convention that serve to indicate what the theoretical terms shall mean and propositions which express empirical assertions by means of interpreted terms and are subject to confirmation or withdrawal of confirmation by empirical test". (Hempel, 1974, p. 82). We shall not go into the fact here that this conception is found not only in Hempel, Carnap, Reichenbach and Stegmüller, but also (in irrelevant variations) in e.g., Popper, Lakatos, Feyerabend and others.

The closeness of this conception, according to which terminological definitions are not to be separated from the testing of the validity of the propositions formulated with these terms, to (older) conceptions, according to which the terms of an axiomatic theory have meaning only in the context of the whole theory, is notorious. Already G. Frege reckoned with the possibility that the terms of Newtonian mechanics would not be definable singly in a hierarchical arrangement, but only all together by means of a complete system of definitions and axioms. (Frege, 1891, cf. also Janich 1975) This view finally finds its most radical expression in Hilbert's formalistic view of geometry (which view Frege, moreover, took to be miscarried (Kambartel, 1975). In the field of mathematics it formed the model for viewing theories of empirical sciences as interpreted formal systems; the empirical verification of such an interpretation is offered then, in fact, as a guideline for the interpretation which lends meaning to the expressions of the formal theory.

The ostensible inseparability of terminological definitions and empirical testing has an effective presupposition, and, in spite of the newly obtained distance from the analysis of scientific language, it is one that is still tacitly carried along: All discussions about validity – or confirmation problems of physical theories, including the variants of a 'non-statement view' (Stegmüller, 1973, p. 12 f.) – proceed from the fact that physics in its present form, in principle yields valid results. Definition problems, validity criteria and even the extreme questions of how one can understand physics beyond of its theories are not really discussed under the risk that this discussion might flow into a critique of physics. Present theories are taken as confirmed, the actual practice of the physicist as successful. With this, however, the undertaking begun with the critical intention of understanding physics as it is acquires a special dependence upon physics: The results of the analytic theory of science cannot be understood if their preliminary judgment that existing physics is the best possible at present is not accepted. In particular, the results of the analysis of the technical language of physics become dependent on the assumption that the technical sciences are correct in principle. The preliminary judgment – one which still stands squarely in the tradition of logical positivism – in favor of physics as the simple exemplar of an exact empirical science and in favor of the decision resulting from it to make definition problems generally dependent on validity problems, points to the *affirmative character* of this direction in the theory of science.

By the designation 'affirmative' we shall not assert that it is the declared intention of the affirmative theory of science to corroborate additionally the system of physics. Rather this predicate is meant to indicate that the affirmative basic attitude (only visible in the analysis of the indispensable presuppositions of this direction in the theory of science) leads to the fact that the theory of science, notwithstanding its otherwise declared intentions, actually assumes a role of confirming physics. For from a mere analysis of physical terminology there can result no objections to physics, because rules of word usage have to be determined by the validity of physical theories. There would be, for the case that a terminology exhibits inconsistencies; but the requirement of logical consistency of theories is generally recognized in the technical sciences. Consequently questions of criteria for the verification of validity of these theories or of an understanding of physics merely take over the task of pointing out, after the fact, in what sense existing physics is recognized. As is well-known, the stock of procedures actually employed by physicists did not develop in a planned way; rather it developed 'historically', i.e., in a mainly evolutionary way. Thus even today physicists as a rule give no information about what validity or verification criteria are used for formulating or verifying their theories; they leave this to the common sense of professional colleagues and practice shows that only in exceptional cases do difficulties result from this. Therefore, everywhere where reference to the consensus of experts does not suffice as a validity criterion, theorists of science only contribute an explication of possible grounds for distinguishing features of this consensus. For the assertion that existing physics is the best possible physics, which assertion is above all ungrounded, they furnish arguments after the fact.

This self-chosen restriction of the affirmative theory of science is not maintained inadvertently or unnoticed by its advocates; rather in cases where the theory of science begins to criticize the special sciences it is frequently expressly defended. (Cf. Stegmüller, 1973, pp. 1–64.) Corresponding to the affirmative basic attitude of theoreticians of science is their espousal of the autonomy of the special sciences vis-à-vis philosophy. With this, however, there underlies the affirmative theory of science – in addition to shortcomings to be criticized later – the misconception that the *de post facto nature* of reflections in the theory of science requires the *acceptance* of existing theories. Starting from the fact of the natural sciences, something certainly with which no philosophy of science can dispense, is, however, also possible without a preliminary value judgment in regard to existing science; this appears not to have been seen up to today in the tradition of logical positivism. (To avoid misunderstandings, let it be pointed out that rejecting a preliminary judgment clearly also includes the negative value judgment that the theories of physics are poorly verified or otherwise defective.)

In this book, we likewise advocate starting from actually existing physics, nevertheless without evaluating it in advance – whether positively as a model of scientific rigor and successful research, or negatively as an example for the restriction of a viewpoint or methodology. The advantages of distinguishing philosophical efforts toward an understanding of existing physics from the acceptance of this phyiscs and of doing only the first without yielding to a preliminary decision will be treated in the second chapter. Here we shall present first the difficulties connected with a positive preliminary decision vis-à-vis physics.

If one assigns to the theory of science, among other things, the task of contributing to a better understanding of physics in its respective historical form, then beginning with an analysis of scientific language was not only delineated in the history of the exact sciences as was said above; this approach has also proven itself to be fruitful and successful. As its success, it must be namely of value to know that the actual language used by physicists themselves in theories that have a certain closedness cannot be reconstructed in the sense that the terms of the theory would be completely translatable into a non-theoretical language. Examples of such terms are, for one, all metric concepts in which real coefficients of measure are used, although measurement results can only be rational numbers, and, for another, terms such as 'electron' or the Schrödinger wave function ψ. This success in analyzing and describing the actual usage of long wage among professional scientists is a presupposition for constructively criticizing this usage and finally also for deciding, as theorists of science, to seek for alternative procedures for constructing a scientific language. This search, however, has not been seen as a problem by those theoreticians of science who assume an affirmative basic attitude vis-à-vis physics.

With their self-chosen restriction they do not consider, among other things, the fact that professional scientists are humans who pursue various ends when they do science, more briefly, that they act purposively. However, already with the description of physics that is actually done, one may not presuppose without examination that professional scientists act only in accord with ends immanent in their discipline, i.e., that physics is autonomous in its aims; nor may one presuppose that aims external to the discipline have no effect on the results of physics and consequently also on its linguistic regularization. Even where, in addition to the theory of the language of science, questions of the testing and verification of physical theories have also been discussed and where this discussion has led to the perception that the history of science is, among other things, a social process, (Kuhn, 1970) there has been a failure because of the affirmative basic attitude to comprehend physics as a good-governed activity. It was not seen, and that inclusive of positions following Kuhn, that a new problem presents itself to philosophy of physics precisely when the consensus of professional scientists, e.g., in regards to their technical terminology or their theories, is argumentatively incomplete or circular. Kuhn indeed assigns a critical task to the theory of science, but with that he neither departs from the limitation to a scientific community nor can he formulate concrete proposals for changing physical theories. Consequently the most important deficiencies of the so-called *theory-languages account*, (Janich, Kambartel, Mittelstrass, 1974, pp. 50–54, 90–95) which distinguishes 'theoretical concepts,' have not been resolved up to now.

The chief error in attempting to arrive at the statement of (in the broadest sense) procedures for defining physical terms by means of a logical analysis of the existing technical language alone lies in the overdrawn claim of this account. All terms necessary for the formulation of a theory, so far as they do not belong to logic or mathematics, are reckoned in like manner among the theoretical ones, the use of which depends on corroborating the theory by means of measuring controls. The terminological definition of all words occurring in physical theories is supposed to be empirically dependent (with the mentioned qualification). Here it is overlooked that some terms concern those properties of measuring devices that are presuppositions for any measuring control. Already consideration of the actual linguistic usage of physicists could have taught that some physical terms do not cover natural or naturally lawful states of affairs, but rather technical, i.e., artificial ones. It has always been taken as a consensus of physicists that the reliability of physical results is produced by the repeatability of experiments and that this repeatability consists in the reproducibility of the experimental conditions. Nevertheless this consensus of physicists has not penetrated to the theory of scientific language, otherwise it would have had to be considered there that, in their use, some theoretical terms of physics are defined not by hypothetical statements to be verified empirically; rather they depend on the claim to validity of propositions about individual measurement results. They secure this claim to validity through their reference to the experimentally reproduced properties of instruments.

Clearly it cannot be inferred from the formulas of a physical theory what terms, now, are available with the empirical testing of the theory and what terms, on the contrary, are a presupposition for the theory's being able in general to he tested empirically. This distinction is not effected by distinguishing the syntactic basic concepts – something analytically possible – where these basic concepts are those which, in the framework of the syntax of a particular theory, appear not as explicitly defined, but as 'primitive' terms, yet suffice for the definition of all other terms. The syntax of a theory cannot show whether the primitive terms can be defined so that they have meaning as a designation for the technically or artificially produced properties of instruments and consequently have it independently from the validity of the theory formulated with them. The possibility, however, that physical basic concepts allow themselves to be marked out according to criteria other than syntactic is unappreciated.

Naturally it was, or is, also known to the advocates of the linguistic-analytic theory of science that the significant properties of measuring instruments are artificially produced and that experiments yield experiences under artificial conditions. Therefore one must explain additionally why the distinction of empirical and artificial properties of instruments has acquired no influence on the analysis of physical theories.

The positive preliminary judgment in favor of physics concerns not only the assumption that existing theories are at the moment the best possible, but it also extends itself to the laboratory work of physics: Here the affirmative basic attitude expresses itself in the view that the history of laboratory technology can be forgotten or is at least irrelevant for physical theories, if the presently employed instruments can be described with the presently accepted theories.

Such a neglect of the history of laboratory work found its most extreme expression in Peter Mittelstaedt. His view, borrowed from C. F. v. Weizsäcker, of the circular structure of the growth of physical knowledge still has there the philosophical background of interpreting the world as a natural scientist, as it were, by questioning behind the beginning of existing theories, after the analogy of a hermeneutical circle. In Mittelstaedt the reactions of scientific, experimental experiences upon everyday experience merely explicate the conception that classical physics is to be understood as a limiting case of relativistic and quantum physics and they contribute nothing to the foundation of modern phyiscs. Due to the logical circularity of his proposal neither the history of the art of experimentation nor its systematic foundation can be understood. Cf. Mittelstaedt, 1972.

The language of measurement results, which are indeed always also propositions about measuring instruments, must remain consistent with the language of theory. In a metaphor stemming from Wittgenstein, the history of the art of constructing measuring instruments can be forgotten, just as one can throw away a ladder once one has climbed a roof with it. It was believed possible to forget, along with the history of the development of measuring instruments, its systematic implications as well, namely that certain improvements in instruments presuppose other measurements, not only historically but also methodologically; this was believed possible because linguistic analysis, which was after the fact vis-à-vis the history of physics, ensued subsequently to the reinterpretation of the basic concepts of classical phyiscs. The legitimate criticism of Poincaré, Minkowski and Einstein of the conception basis of classical physics, as is well-known, has also gone so far as to reinterpret the measurement results supporting relativity theory in respect to the properties of measuring instruments implied in them. Thus just as the emphasis on linguistic analysis in the theory of science was already delineated by the history of the exact sciences, so also grounds in the history of science have been delineated which have supplanted the aspect of artificial production of crucial properties of measuring instruments in favor of a depost facto description in the words of the theory. Later we will once more have to go into the fact that, especially in the analysis of the time concept of modern physics, it becomes clear how much this analysis is stamped by the historical accident that it has to follow the methodological shift of physics due to Einstein's relativity theory.

Besides this historical reason, there is yet a further one for not having distinguished empirical properties of measuring instrument from artificial ones: It should not be asserted here that the choice of physics as a model for scientific nature (Wissenschaftlichkeit) was simply naive. The methodological maxim of restricting a science of nature to the observable has an enlightening character compared with the taint of Newton's mechanics by theological influences. Keeping physics free from such traditional remnants, which prove themselves to be assertions that are not rigorously demonstrable in argument, would certainly also be generally agreed to today. This agreement frequently expresses itself in the opinion that physics owes its success directly to its detachment from a traditional philosophy. Unfortunately, already with Einstein, however, since the explication of his special theory of relativity by scholars such as Schlick and Reichenbach, the enlightened impetus extended and intensified to generally suspecting everything of metaphysics[3] that is not the *result* of an empirical science. Any assertion of the sort that there indeed could generally be basic concepts of physics which in their definitions exempt themselves from an *empirical-scientific* revision is also met today by such a suspicion. This is expressed above all in objections from the tradition of critical rationalism and is burdened by the assumption of some critics that Dingler's fundamentalism is continued in protophysics. The misconception that asserting the existence of a presupposition for any possible physics can, scientifically, only be unverifiably valid is too obvious. The old, and here advocated, aim of restricting physics to rigorously provable statements has consequently led to a suspicion, itself unfounded, of efforts such as the ones presented here.

It is, namely, not just 'metaphysically' (in the perjorative sense) possible to speak about the general presuppositions of any empirical physics, because with this, particular minimal demands for the validity of this physics are always implied. In Chapter II we discuss at great length the fact that already the methodological decision to carry out experiments with the help of *measuring instruments* and to keep *measuring results intersubjectively testable*, i.e., in the strict sense to carry on an experimental science, requires those efforts that lead to the intersubjective executability of measurement results. Thus, in light of the claim of the validity of physics, we can argue strictly for the fact that every physics that builds upon measurement results presupposes, in regards to the measuring instruments, properties that are logically exactly definable; and if this presupposition is not itself to be unverifiable, it must be validated by technical arrangements. For this, language is indeed already necessary and certain experience must also precede such technical efforts; but for this neither the terminology of physics nor its empirical knowledge is a precondition. The artisan's language and everyday knowledge suffice.

The old suspicion of metaphysics on the part of the empiricist tradition, consequently, entailed the fact that the distinction of everyday knowledge and scientific knowledge, and finally even the distinction of ordinary language and scientific language, remained closed to the empiristic and analytic philosophy of physics, insofar as it is precisely *the claim* of intersubjective verifiability of physical propositions in opposition to empirical sentences in daily life that supports this distinction.

2.a. *M. Bunge's Affirmative Protophysics.*

The preceding critique of the affirmative basic attitude of the theory of science that proceeds analytically [in the sense of linguistic analysis] was written, above all, for the philosophers of the Vienna Circle and its tradition. Things are somewhat different in the works of Mario Bunge, who emphasizes the necessity of research into physical foundations and simultaneously asserts that in this there are, practically speaking, no results (which may also be understood as a rebuff to the results of the authors cited above) (Burge, 1967, p. 5). Bunge pushes to an extreme the affirmative basic attitude towards physics. It is not only that physicists alone must decide what physics is and what not, not only that each new development of physics requires a new foundational inquiry, but even foundational research of already existing theories is an endless undertaking (Bunge, 1967, p. 197).

Although Bunge remains critical towards other preliminary decisions which representatives of the Vienna Circle accept, he himself makes an extremely radical preliminary decision which Carnap, Hempel or even Popper have not accepted: the circumstance that foundational research is an activity secondary to physics (Bunge, 1967, p. 1) confines Bunge to the position that foundational research has *only theories* as its subject matter. An extreme example of this is the assertion occurring in his critique of operationalism that already each operation, say a particular measurement, can only be interpreted with the help of a theory (Bunge, 1967, p. 28), i.e. that even elementary operations presuppose theories (Bunge, 1967, p. 27). Consequently, according to Bunge himself, foundational research, not in its analytic, but in its constructive parts (Bunge expressly assigns numerous constructive tasks to the foundational research of physics) becomes a metatheory of theories, which indeed claims to stand closer to physics than do the works of Carnap, Hempel and others, but which itself, is concerned only with theoretical physics.

Where Bunge requires protophysics,[4] and that with the argument that trivialities may remain unnoticed only as long as they are under control (Bunge, 1967, p. 86), it must be objected to him that he does not have the physicist-triviality under control, according to which some operations in experimental physics must not be interpreted after the fact and by no means theory-dependently, because they are sufficiently defined in characteristic components by the intention of the acting experimental physicist. This objection would no longer hold only if ordinary speech about intentions, merely as long as it is terminologically and logically rigorous, were to be counted among the 'theories'. Then, indeed, any reflective speech would become a theory and Bunge's assertion that no theories can manage without theoretical (i.e., operationally uninterpretable) expressions would become an assertion about ordinary language that has not been established by Bunge himself. What the consequences are for theories of measurement or especially for a theory of time measurement, when foundational research exclusively investigates theories, will have to be shown in the following.

3. THE AFFIRMATIVE THEORY OF MEASUREMENT

The criticism, presented in general in the previous section, of the affirmative basic attitude of a distinctly analytic theory of science will now be sharpended for theories of measurement and the consequences of this attitude for the understanding of physical basic concepts will be shown. One can presuppose as non-controversial that basic terms, at least within certain theories, can be distinguished syntactically. Existing theories of physics have for the most part reached a degree of syntactic regimentation which permits one to give respectively for a closed theory its basic terms. These expressions, occasionally also called primitives or primitive terms, are not defined explicitly within the theory, but permit (possibly calling on sentences with an assertory character) the definition of all remaining terms of the theory. The syntactic distinguishing need not be unique, i.e., there can also be several groups of primitives in one and the same theory, which groups suffice to define the respective remnants. Whether now unique or not, the syntactic distinguishing of basic terms always ensues relative to a completely definite theory and does not presuppose that in general 'basic concepts' of *physics proper*, i.e., physics independently of the momentary form of its theories, could be given. Therefore, in the following, we shall use the expression *syntactic basic terms* for this.

As is well-known, it is the syntactic basic terms of mechanics which have experienced the most detailed discussion in the philosophy of physics, i.e., the concepts of length, duration and mass or force. Although the problem of defining these concepts was widely abandoned to the language theory account – according to which only by interpreting whole theories such as geometry or a mechanics can it be known what 'length', 'duration' or 'mass' mean – a 'theory of metrical concepts' (Above all in: Hempel, 1952) was able to be developed in which the logical properties of equality and order relations are established. Historically, questions of this type come from geometry and, in connection with the *Ausdehnungslehre* of H. Grassmann, were first comprehensively carried over to the explication of measurement by H. v. Helmholtz. (Helmholtz, 1887, cited from Hörz and Wollgast, 1971) Helmholtz's well-known account already, in spite of later technical and linguistic-philosophical improvements by some representatives of the Vienna Circle, contains the core of the theory of metrical concepts still recognized today.

Helmholtz starts from the question of how arithmetic is to be applied to physical objects, whereby he is critical enough to emphasize that his 'empiristic theory' directed against Kant "must also justify itself via the origin of the arithmetic axioms." (Helmholtz, 1971, p. 301) Measurement is for Helmholtz the determination of a coefficient of measure: Of much greater significance and more extensive application (e.g., than the counting of unequal parts of a straight line, duration, etc.) is the counting of equal objects. Such objects, which in some definite relation or other are equal and are counted, we call the units of counting, the quantity of the same we designate as a concrete number (*benannte Zahl*), the particular type of unit which comprises them, the concretization (*Benennung*) of the number. (Helmholtz, 1971, p. 318) Therewith two problems present themselves: (1) How are equality and assembling of physical objects regulated so that these operations possess the same logical properties as equality and addition in arithmetic? (2) How are the units determined?

Helmholtz's answer to (1) proves surprising if one compares it with corresponding ones of Carnap and Hempel, for, in spite of Helmholtz's empiristic position, it is conventionalistic. Symmetry and comparativity of equality as well as commutativity and associativity of the assembling operation serve thereafter with extensive, i.e., additive quantities like length or weight, to characterize a type of quantity; for equality Helmholtz argues explicitly, using the example of weight comparison on the balance, that symmetry and comparativity are criteria for the correct, disturbance-free functioning of the balance. When two bodies *a* and *b* are in equilibrium on a balance and are then exchanged, "the balance must remain in equilibrium. That is the well-known test of whether the balance is correct". (Helmholtz, 1971, p. 322) With this, in Helmholtz, there arises the characterization of operations on disturbance-free ('correct') instruments. Moreover, Helmholtz emphasizes that non-additive ('intensive') quantities (like temperature) are methodologically dependent, in their definition, on those extensive quantities: "On the other hand, however, it follows from the given derivation that one must have first formed additive quantities before one can define coefficients. For the equation which expresses a natural law can give the valuation of a coefficient only if all quantities occurring in it are already defined as quantities. Thus the definition of additive quantities must always take precedence before one can find the values of non-additive ones" (Helmholtz,1971, p. 331).

As shown in the following, the insight into the conventional character – namely that character defining the undisturbedness of measuring instruments – of the logical properties of equality and assembling of physical objects was again abandoned in the affirmative theory of science of Carnap and Hempel which concentrates on linguistic analysis. In the meantime, the answer given to (2) by Helmholtz will be continued.

The unit of counting required to ascertain a ration number (coefficient of measure) is either 'arbitrarily chosen' or 'given by the instrument used" (Helmholtz, 1971, p. 319). Only tracing such *ad hoc* chosen counting units back to naturally given ("only to be exhibited in certain natural bodies (weights, measures), certain natural events (day, pendulum oscillation)") leads to "absolute units of physics". This is an extremely empiricistic view, because it rates the natural givenness of units more highly than their availability at any time through their being able to be reproduced at any time. Among the followers of Helmholtz this overrating leads in the extreme case to the view that certain quantities can be measured only because we can find naturally given units for the measuring. Followers of this view thereby prove themselves to be not only poor students of Euclid, who, in the procedure of alternate subtraction (*Wechselwegnahme ἀνταφαίρεσις*) demonstrated the determination of rational coefficients of measures without the previous fixation of a unit,[5] but also undeserved heirs to an empirical physics which owes its success to the art of 'approaching nature' in the experiment with measuring instruments that are artificially reproducible in their crucial properties – 'approaching it,' indeed to be taught by it, but not in the character of a student who listens to everything that the teacher chooses to say, but of an appointed judge who compels the witnesses to answer questions which he has himself formulated (Kant, 1787, 8 XIII).

Naturally no objections against the use of general physical units like meter, second, pound, degree centigrade, etc. shall be raised here. Besides an honorable tradition, there are also good reasons for reducing measurement, whose result should indeed be a (rational) coefficient of measure, to the elementary activity of counting. But apparently with all such reduction attempts too much is borrowed from arithmetic; this borrowing, indeed, indulges the tradition since Descartes and since the arithmetization of geometry, but it also simply passes over the preceding characterization of the objects counted. Thus, admittedly, the operation of assembling pieces of one size suggests counting with natural numbers, and due to the reversal of the assembling, with whole, i.e., also negative numbers; and the division of pieces until a common measure for two given pieces is obtained leads to rational coefficients of measure. But already the leap from the rational to the irrational numbers can no longer be solved within the bounds of a program of explaining physical quantitites through the isomorphism of measuring operations and the mathematical description of the same. Here, approaches to reducing the terms of physical theories to observation language founder; here there are formulated *ad hoc* principles for commensurability (Hempel, 1952, p. 67) that are formally incompatible with the mathematical apparatus of physics and consequently are abandoned in favor of ('partial') interpretations that are possible only in the framework of whole theories.[6] Obviously the existence of a finished construction of the system of real numbers attempted to be carried out with the help of measuring operations operates as a hindrance here; it hinders, namely, the attempt to grant measuring operations a primary status compared with arithmetic and to undertake the differentiation of number systems themselves firstly by the differentiation of measuring, i.e., to (re-) construct the system of rational and real numbers in a construction of the theory. How this construction might look, can be gathered e.g., from Lorenzen, 1965.

The copying of the historically older construction of our number system in a theory of measurement (but not, thereby, in the practice of measurement!) is repeated once more in the affirmative theory of physical concept formation. Corresponding to the modern set-theoretic characterization of numbers since Dedekind and Weierstrass, there has been worked out by Carnap and Hempel a theory of a progressive construction of concepts which understands quantitative (or metric) concepts by sharpening comparative (or topological) ones, which for their part stand above classificatory concepts. Here the ineradicable topos of the progress of empirical science that begins with the accidental and then systematic collection of individual experiences, and via their ordering leads to quantification[7] has undergone a formally exact elaboration. According to this theory a group of pair-wise contrary predicates permits a partition of an object domain into disjunct classes. The definition of an identity and a predecessor relation defines for these objects a quasi-ordering, which, by means of further principles, can be sharpened to a metric. This progressive construction yields the well-known construction, one which also belongs to the basic knowledge of e.g., modern psychology, of types of scales: nominal scale, ordinal scale, rational scale, interval scale.[8] The formal criteria employed for this progressive construction are sufficient – and this, relative to an analytic interest, can be considered an advantage – sufficient to distinguish physical quantities with and without absolute zero, additive and non-additive, fundamental and non-fundamental quantities and still others and finally even to describe the transition from the classical to the relativistic time concept as a change from an extensive to an intensive one (cf. Stegmüller, 1970, Chapters I and II). The benefit of these results of the analysis of physical concept formation for an understanding of modern physics stands beyond doubt. The formal elaboration and detailed explication of this theory of concepts and the explanation implied therein, however, also allow as to state the deficiencies of the analytic account:

(1) It is highly doubtful whether in the distinction 'classificatory – comparative – quantitative' there is at all present a hierarchical construction of concept types. The view carefully presented, e.g., by Stegmüller that in the "early phases of language" there prevails "the simplest category" of the qualitative concepts, "presumably exclusively" (Stegmüller, 1970, p. 19) is neither reasonable nor supported by the ontogenetic or the phylogenetic designation of "early phase". One-place as well as two-place predicates, regardless of whether the latter are symmetrical or not, are acquired and employed in identical processes. It is not to be seen how, say, the acquisition of relators for designating family relationships should begin with one-place predicates ("*X* is father") and then, by adding to them, lead to two-place ones ("*X* is the father of *Y*"). The assertion of a higher developmental stage of a language holds good, thus, only for the application of arithmetic to polar contrary predicates such as long-short, hot-cold, etc.

Carnap, Hempel and Stegmüller argue for their conception of the hierarchy (Stufenaufbau) of scientific categories by using the advantages of metric concepts over the classificatory and comparative ones (cf., among others, Hempel, 1952, pp. 54–58). Thus, for example, Hempel, in a comparison of the wind-strength scale of Beaufort with the indication of wind-speed in meters per second, sees the advantages of the latter determination of wind strength in the fact that this would make possible finer, in principle even infinitely, many readings (in comparison to merely 12 wind strengths), that it directly shows the position of a speed value in the scale (in the Beaufort scale one must learn by heart that a 'moderate breeze' comes before a 'fresh breeze' in the scale) and finally in the applicability of mathematics.

Surely there are contexts in which these arguments hold, in the case of the given example, say, meteorology. But the impression suggested thereby that metric scales should be constructed successively (and precisely therefore to advantage), beginning with classificatory ones, by means of step-by-step exact definition is false. The Beaufort wind-strength scale is not a preliminary form of a scientific scale of wind-speed – rather, now as ever, it has practical meaning in sea travel or for yachtsmen, where it is neither a question of finer data nor is a measurement of wind speed possible (e.g., due to unreasonable technical expense or due to superimposition of travel wind on the true wind speed, etc.). As in this case, so in all cases, the context of purpose plays the deciding role. The greater possibility of differentiation of metric scales and the applicability of mathematics are advantages only there where, in the light of the technical manipulability of events, the mathematical formulation of laws for metric quantities is desired generally.

Connected to the praise for the advantages of metric vis-à-vis less sharp concepts are common misunderstandings in regards to the criteria of scientific nature. They are found chiefly in the social sciences where application of mathematics is mistaken for intersubjective verifiability of propositions. Some theorists of science, however, also seem to be of this opinion. ["Here it is shown clearly that with the process of metrization a process of objectification, i.e., of obtaining common intersubjective results, is introduced at the same time." Stegmüller, 1979, p. 64]. Against this it remains to be remembered that many scales lower than metrical ones thoroughly fulfill their purpose, that sharpening lower scales to metric scales frequently produces only a pseudo-scientific nature, and that even propositions with metric concepts are always only intersubjectively testable to the extent that the definition of these concepts is testable. For this, in many concrete cases, e.g., in metric of lengths, it is precisely those ways that are necessary or at least expedient that do not follow the construction classificatory – comparative – quantitative, but which, e.g., in the metrization of lengths, proceed via the concepts 'straight, congruent, continually divisible'. It is consequently of little reasonableness to state in general the advantages of refining concepts (in the sense of ascending in the hierarchy of categories), because each stage of this 'hierarchy' holds out – according to the context of purposes – an independent, significant linguistic possibility that is perhaps misunderstood as a preliminary form of 'higher' stages.

With this we shall not generally dispute the fact that, subsequently to an existing vocabulary exhibiting various categories, criteria may be given which give rise to the mentioned hierarchy; but these criteria as a rule have little to do with a linguistic construction that follows in methodological sequence.

(2) To the rejection of the hierarchical construction of conceptual and scalar categories these corresponds the rejection of the view contained in it of the inductive procedure of the empirical sciences. No path leads from the

sorting of sticks according to long and short, via ordering them according to longer and shorter, to the determination of their rational length proportions and finally to the empirical verification of e.g., Pythagoras's theorem. For (to mention only one objection to this 'construction') stipulation of an operation for combining sticks according to the logical properties of arithmetic addition overlooks the fact that the combined rods not only must be laid in a straight line one after the other (something that shall be guaranteed by the isomorphism to addition), but that they themselves must already be straight. In short, the definition of 'straight' by additivity properties of (straight!) sections, or equivalently through the condition of being the shortest connection of two (end-) points is obviously incomplete or circular.[9]

(3) The central objection, however, is that the intended task of reducing measuring to counting is not carried out. The reason may be briefly characterized by the fact that one can, indeed, count naturally given objects, but that comparisons of sizes according to the analytic theory of comparative and quantitative concepts cannot be done on natural objects, but only on artificially prepared objects.

We shall show this in detail with the example of the balance already chosen by Helmholtz.[10] There is needed, for a comparative as well as for a quantitative concept of weight, an equality relation for bodies relative to their weight, a relation that is operatively defined by the horizontal position of the balance. Symmetry and comparativity must hold for this equality, i.e., it must be an equivalence. With this there results the problem of what status the proposition that equality of weight is an equivalence relation has. There is agreement that it is non-analytic. For the reasons already mentioned above, namely that (a) real numbers should be applicable to the gravitational mass and that (b) the analytic theory of science must also comprehend the reinterpretation of the basic concepts of classical physics through relativistic physics, this sentence, as is well-known, is taken to be empirical. With that, the affirmative theory of science has led to an empirically-analytically unresolvable dilemma: There is namely no possibility of empirically verifying this supposedly empirical sentence. It would be only possible on a balance that functions free of disturbances; and as criteria for this freedom from disturbances there are again available no other definitions themselves than that equality of weight is an equivalence. If a balance is free from disturbances, then the equality of weight is an equivalence relation. If one adheres here to the dogma that all sentences are either analytic or empirical, then only the possibility of now regarding this sentence as analytic remains. If, namely, the freedom from disturbances of the balance is defined, as in Helmholtz, precisely by means of the fact that equality of weight is an equivalence, then the sentence is a tautology of the form $A \rightarrow A$. With that it would lose any significance for an empirical science. If one, however, departs from this dogma, then the sentence "Equality of weight is an equivalence" can be understood as a command to manufacturers and users of balances to so regulate balances that equality of weight becomes an equivalence.

This example should hold for all properties enumerated in the theory of comparative and quantitative concepts: They are to be produced artificially, namely technically, in the measuring procedure or on measuring instruments.

(4) This objection, that in an affirmative basic attitude one believes scientifically confirmed what, correctly viewed, consists in the human intelligence of making empirical science possible, has further implications. The focussing of analytic theory of science on linguistic analysis and induction or validity problems has also developed the theory of measurement merely as linguistic analysis. Pointedly stated, there is no analytic theory of measurement, rather, as an inadequate substitute for it, merely a theory of quantitative concepts. This theory lacks namely the viewpoint that measurement is a planned activity that can be understood only from the starting point of its intentions. Intentions that are constitutive for properties of measuring instruments and therewith also for the logical properties of measurement results can indeed fail but they can never be refuted by measurement. Consequently the logical properties of the equality or inequality relation that are resorted to to characterize metric concepts are not empirical sentences, indeed not even declarative ones, but rules for the production and use of measuring instruments.

Now the suspicion is obvious that this critique would remain external to the analytic theory of science, i.e., not concern it, because the analytic theory stands in a tradition which deliberately refrains from taking into consideration 'psychic' components like, say, the intentions of experimental physicists. It, however, becomes immediately evident that the case is otherwise. The confounding of empirical and intentionally produced properties of measuring instruments that arises from the affirmative basic attitude results in the total absense of a theory of disturbances. In this it is not disputed by anyone that measuring instruments can be disturbed and that a good part of the art of experimenting consists precisely in avoiding such disturbances. One might think of the trivial example where a clock stops because the drive source is exhausted. Naturally we can state with physical laws why the clock does not run any more. The stopping still of the clock is therefore empirically explainable. But that the clock then is no longer useful for time measurement, because there exists a particular disturbance in its operation is no empirical fact, it follows rather from our expectations of clocks. The lack of possibilities of distinguishing empirically and artificially produced properties of measuring instruments thus leads to the absence of the distinction between a disturbance and naturally lawful behavior. With that it obstructs access to the question, which is central for the understanding of physics, of what we can know about nature from experience and what, as a historical cultural accomplishment of humanity, is our proven ability to pose questions to nature.

It appears as though Carnap takes the occurrence of disturbances to be possible only with the choice of units (e.g., of length, of time). He mentions, as an example, that after choosing a particular standard measuring stick one would have to additionally stipulate by convention that his stick maintains a particular temperature, otherwise when this stick is heated a shrinking of the world would manifest itself and physical laws of high irregularity would be the result. This example impressively proves the disadvantages of the analytic theory of metric concepts. It is a generally recognized requirement that physical laws be so formulated that they are invariant relative to a conventional choice of units. This, however, also suggests a characterization of metric concepts in which the units play no role, i.e., a characterization in which the unique determination of proportions is guaranteed. Here, in Carnap, the old empiricist conception that measuring must be reduced to counting identical objects and that the criteria for identity are themselves subject to modification by new measuring experiences shows through.

It is thus not surprising if the affirmative theory of science has found no declared position toward conventionalism. In general it is not disputed that, e.g., in the construction of a physical technical language, conventions are always also encountered. But in general one also turns against a conventionalism which stands under the suspicion of wishing to immunize physical theories against being able to founder on experience. Here Popper is to be cited in the first place; he indeed entirely points out conventional components of his own falsificationist theory, but also, almost pathetically, emphasizes how much all results of our intellectual efforts must be available at any time. Regrettably Popper names, among the conventionalists, Poincaré and Dingler in one breath, which is certainly improper (Cf. Diederich, 1974, where on one hand the differene between Poincare and Dingler is clearly elaborated and on the other Popper's critique of conventionalism is recognized as, in its subject matter, being chiefly directed against Dingler. Also attitudes of early logical empiricism towards conventionalism are reported there.) The problem of what is conventional in a physical law like the Newtonian law of gravity, and of what can founder on experience cannot be treated here. It can, however, be explained to what extent Poincaré's conventionalism was misunderstood by the philosophers of the Vienna Circle as well as by Popper. Poincaré, among other things, had asserted the conventional character of geometry. For the case that geometry is understood as the system of demands on the undisturbedness of measurement of lengths and angles, it is thereby not in fact a matter of theorems that can be refuted by measurement. With sentences about time measurement the case is analogous. Since, however, relativistic physics has carried out a reinterpretation precisely of the basic concepts of space and time as well and since the affirmative posture must take this circumstance into account, neither Carnap, Reichenbach and Hempel, nor Popper could recognize a difference, e.g., between the Pythagorean theorem (as a sample sentence for a criterion of length measurement) and the Newtonian law of gravity (as a sample empirical sentence.)

4. AFFIRMATIVE EXPLANATIONS OF THE CHOICE OF THE TIME STANDARD

After having discussed in the previous section the problem of the affirmative basic attitude leading to the development of theories of measurement solely as analytic theories of scientific language and, above all, of metric concepts, we must still discuss the problems of time measurement in particular. In accordance with the results of the above-cited view, to do this we must still give standard motions that lead to time units or otherwise correspond to relationships formulated in theories of concepts. It is thus here more precisely a question of what standard events allow one to more closely characterize time measurement on a scale.

The proposals presented in the available literature may first be divided according to whether they fundamentally metricize time, i.e., in particular manage without geometric propositions about standard events, or whether they apply a metric of length or angle to particular events such as earth rotation or inertial translations. This division does not coincide with others in the literature, so that the proposals discussed in the following are not to be understood as a number of disjunct possibilities.

Many authors emphasize the possibility of a fundamental metrization of time, some even hold a derived metrization with the aid of geometry to be an historical misfortune. G. J. Whitrow must be so understood when he complains about the fixation of an almost 2000-year long history of physics on the model of Euclid and Archimedes (instead of on Aristotle). (Cf. Whitrow, 1961.) In any case, with fundamental metrization the view criticized above comes to bear, namely the view that the definition of a physical quantity must begin by stipulating a unit and with that construct a (when possible, metric) scale. Accordingly this definition begins by designating an event which is supposed to form this time unit. The historical fact that men undertook initial temporal definitions with the natural counting units day, moon phase and year suggests natural standard events. Unfortunately the label 'natural' is, however, already handled here very unclearly. For one, a natural event can be meant in contradistinction to a technically realized one (a), thus e.g., the rotation of the earth against the fixed star heaven in contradistinction to the oscillation of the balance of a pocket watch; or 'natural' can signify 'naturally lawful' (b), as one knows, say, the physical condtiions for the constant oscillation of a pendulum or quartz crystal; and finally (c) 'natural' sometimes also signifies that certain events, namely precisely the periodic ones, are suited to be time measurement events because they exhibit 'according to their nature' a precisely designated start and a precisely designated end. This last possibility is terminologically misleading. What would be better said here is that periodic events are suited per definitionem to designate a time segment considered as a counting unit. The existence of distinguished repeating states can be an argument for the preference of periodic motions like pendulum oscillations over motions of constant velocity only where otherwise (e.g., geometrically) no motion phases can be fixed. It is not clear that this difficulty can appear at all, since certainly with every uniform motion, e.g., the earth rotation, a reference system for the motion must be given to which then a geometric scale can also refer.

(a) The selection of natural events (in contradistinction to those technically produced in machines (clocks)) raises a problem: There is an enormously great number of natural events that form no class in the sense that they would have (at least temporarily) a constant running or frequency ratio. Among the natural must be counted, namely, all organic processes. It would be highly problematic to wish to undertake already here the marking out of 'inanimate natural processes' such as earth rotation or the oscillations in the intensity of the light of pulsars. For where this designation intends to be terminologically exact it must resort to a natural science that already possesses time-measurement. There, then, those motions would be viewed as inanimate that can be described physically so completely as, say, inertial motions or oscillations of an oscillator, that the assumption of a living thing responsible for the temporal course of the motion (as with the beat of a bird's wings) is superfluous. It is thus also wrong to assert that there could be a single class of natural motions that run constant relative to one another and to think thereby, say, of astronomical interial motions. (The argument that only one class of periodic motions exists seems to be characteristic for empiricist programs of metrization. Cf. Essler, 1971, p. 75.) Natural motions in the sense of 'not determined technically or by a living things' belong, due to their definitional presuppositions, to those, according to (b), 'determined by natural laws', or they are not at all determined. Thus where, e.g., the earth rotation is referred to as a standard process for the time unit, this happens either with the implicit presupposition that this is an extremely disturbance-free inertial motion, i.e., with an implicit reference to physical laws, or it happens for merely historical reasons. In the latter case, it is indeed forgotten to consider whether the historical reasons for reference to the earth rotation (such as, say, navigational problems) could also be valid for today's physicist.

The by far most frequent case of a selection of 'natural standard processes' for chronometry concerns the designation in accord with physical laws and includes technically produced motions. The problems of circularity appearing thereby arise likewise for periodic as well as for uniform motions, i.e., manifest themselves both in the case of fundamental and in the case of derived metrization and shall thus be treated in common.

Wherever the selection of definite standard processes for time measurement is argued for, it is a matter of simple logically circular inferences; for the selection of standard processes must have already been made in order to obtain arguments for it from physical theories based on time-measurement. This danger of the logical circle conceals itself frequently unrecognizedly in definitions of the inertial system, where force-free motions are taken as a time-measurement standard, this danger is seen though by most philosophers of physics and they try to avoid it. With inertial motions this obviously has been held to be hopeless, so that in the end the discussion has been limited generally to periodic motions. Here above all Schlick, Carnap, Reichenbach and Hempel pointed out that 'periodic' cannot be defined non-circularly by requiring the recurrence of (within certain physical parameters) identical conditions, for the measurement of these parameters as well as the knowledge that it is thereby a question of crucial circumstances already presuppose chronometry. Therefore a weaker and at first glance non-circular definition of 'periodic' was accepted, according to which the recurrence of 'identical' states is required without it being known whether the identity criterion is physically connected with frequency. While there a definition-theoretic foundation is found in Carnap and Hempel for the retreat to the just characterized 'weak periodicity' (Cf. Carnap, 1966, p. 80), according to which periodic motions define the expression 'lasting equally long' and therefore cannot themselves be characterized by the equal duration of periods, Reichenbach argues substantially physically: the "assumption about the operation of certain real mechanisms", i.e., clocks, is in principle not testable (Cf. Reichenbach, 1928, p. 138).

For both, then, the problem likewise presents itself of stating the reasons for the choice actually made by physics (mechanical, quartz, atom, light clocks, etc.). And as far as the literature in the theory of science reaches, all authors fall back upon a surprising and ungrounded thought: The selection takes place according to a simplicity criterion.[11] After the scholastic '*simplicitas sigillum veritatis*' was extended in the philosophy of Mach as a principle of economy to become a crucial maxim of physical theory formation, this step became obvious, above all for the philosophers of the Vienna Circle influenced by Mach. Besides no one would advocate the maxim that one should prefer a more complicated theory to an otherwise equally good but simpler one.

Nevertheless the simplicity argument is untenable for several reasons:

(1) Historically, according to everything that we know about the history of physics, there were never different, mutually irregular processes of time measurement from which then that one was selected that led to the simplest physical theory. If relative deviations were observed in clocks having differing construction principles, then this was always taken to be a sign of disturbances that had to be eliminated technically through further development of the watchmaker art. It is an historical fact that humanity has only developed one sort of time-measurement. This fact cannot be explained by natural laws, because the 'natural laws' requisite for this presuppose the historical fact – restriction to a class of time measurement processes. The assertions of Carnap, Reichenbach, Hempel and others that the simplicity of theories is a selection criterion for time-measurement standards thus miss existing physics and its history.

(2) The simplicity argument is tantamount to replacing the time parameter t in physical laws by a complicated function of an alternative time parameter τ, where τ is thought of as defined by another class of standard processes. As an example for this are mentioned the pulse of the Dalai Lama (Schlick, Hempel) or of the [present] American president (Carnap). Here, then, a worse, because more complicated, alternative to existing theories is constructed after the fact. The possibility of doing this in a logically consistent

way is then evaluated as proof that present theories, with the underlying processes of time-measurement, are the simplest. Demonstrating that existing conditions are the best possible by inventing worse alternatives corresponds indeed to approved principles of a conservative life style, but as a criterion of scientific proof this procedure is out of the question. The simplicity criterion could only then be taken in earnest, in respect to the art of measuring (not only of time), if there were actually alternatives in physical theory formation that were based upon different classes of measurement standards.

(3) The simplicity argument for selecting a time-measurement standard should avoid narrow circles such as those of defining 'periodic' by 'equally long' or testing the operation of clocks with clocks. The path chosen for this nevertheless proves itself to be circular under closer consideration. Namely it merely makes use, as pointed out in (2), of the transcription of recognized physical laws to a new time parameter τ. The possibility that completely new 'natural laws' would be found on the basis of a time-measurement that is irregular vis-à-vis ours is not considered. With that, the simplicity argument reduces itself to the following circle: One selects those time-measurement standards that lead to the simplest theories. The simplest theories are the accepted, present ones. The present theories, however, rest precisely upon the chosen time standards and not upon those that are irregular to them.

With this the criterion of allowing the simplest possible theories has been demonstrated to be unfit for the selection of standard events for time-measurements. Other, possibly non-circular arguments for the choice of particular processes constituting a time unit are lacking. What above was asserted wholesale for the affirmative theory of measurement proves true here: The theory of time-measurement as an answer to the question as to how particular physical terms can be operationally defined by measuring procedures atrophies to a linguistic-analytic description. The logical characteristics of a metric time concept remain thereby insignificant insofar as neither their status (empirical propositions or those of other standing) has been ascertained nor their connection with measuring actions explained. Without criteria for the selection of time-measurement standards, which criteria can be given non-circularly, any theory of a metric time concept hangs as it were in the air.

With the example of time-measurement, our affirmative basic attitude shows itself once more particularly clearly: From the endeavor to understand physics as it is, we have sought for criteria for selecting measurement standards which lead precisely to present physics. From this, then, affirmative arguments resulted in that the supposition that existing physics is the best possible at present finally was deduced, indirectly via measurement criteria, as a result from these criteria. The affirmative theory of time-measurement shows particularly clearly that its deficiencies are caused by the affirmative basic attitude: The decision to metricize time fundamentally is, on one hand, the source of the difficulties pointed out, because not once yet have the geometric criteria (that are within reach of every watchmaker) for the selection of standard events been taken into consideration; on the other hand, it rests upon the after-the-factness of an analysis of special relativity theory. Were it still an unfounded 'ontological' premise of classical physics that physics has to do with space, time and matter, then the time concept in relativity theory would become, as is well-known, the starting point for the criticism of the classical basic concepts. The discussion of the temporal sequence of spatially separated events and the attendant possibility of reducing a metric of length to a metric of time through the assumption of a designated signal velocity in fact suggested for affirmative accounts a foundamental metrization of time. With that, indeed, the Einsteinian program of restricting physics to the observable was widely overshot: In the end, all observations were reckoned among those that constitute experiences, i.e., even those observations that concern artificially produced properties of measuring instruments. Let us quote Reichenbach here as a representative of many others: "Two clocks stand next to each other and one observes that the beginning as well as the end of their periods coincide; if one now waits further, then one observes that again the ends of the periods coincide. Experience thus teaches: Two clocks standing side-by-side, that have equal periods once, always have equal periods" (Cf. Reichenbach, 1928, p. 138). It is nevertheless misleading to count the tautological sentence that clocks are clocks among empirical sentences.

The affirmative basic attitude thus has overlooked, as a consequence of its being subsequent to relativistic physics, the circumstance that physics rests on *apparative experience*. It treats clocks as natural objects, something which goes beyond the practice of physicists and thus can answer at most the question of whether descriptions of measuring instruments are consistent with recognized physical laws. Such consistency considerations, however, do not permit us to reduce the objectivity of results of time-measurement to the reproducibility of standard processes. Pointedly stated, for the affirmative theory of science, the possibility of an objective natural science itself becomes a natural law. According to this view one can then only *believe* in the reproducibility of the crucial properties of clocks. Therefore, we will construct an alternative theory of time-measurement (in Chapter III) upon the attempt to *secure* the reproducibility of time-measurement standards.

To facilitate the reader's confirmation of the fact that the claim that protophysics avoids the deficiencies of present theories will be made good, let us once clearly set forth these deficiencies in conclusion:

(1) The affirmative theory of scientific language confuses problems of definition and problems of validity; so much so that it can finally neither be indicated what physical theories assert, because what their terms mean remains unclear, nor can their validity be seen.

(2) The affirmative theory of science of physics obstructs the understanding of existing physics more than it clarifies it. Namely, it hinders, the distinction of empirically valid sentences from those that hold because of artificially produced properties of measuring instruments.

(3) There is no affirmative theory of measurement. Therefore, disturbances in measuring instruments fail to be characterized, although in the practice of the physicist some measurement results are interpreted as disturbances, others as empirical findings.

(4) No reasons are mentioned for the fact that some physical quantities can or must be metricized fundamentally, others derivatively.

(5) The construction of metric scales corresponding to the affirmative theory of scientific categories fails insofar as it needs empirical sentences about identity and order relations which demonstrably cannot be refuted by experience.

(6) Finally no non-circular selection criteria for standard processes can be given which can make a chronometry possible.

CHAPTER II

ON THE METHOD OF PHYSICS

1. PRELIMINARY REMARKS

The attempt undertaken in this work, to construct a fully founded theory of time-measurement, is to be understood only against the background of a methodology which has been worked out in detail. Therefore, before the conceptual foundation of a chronometry can come into play, we must illuminate the presuppositions of scientific speech to the extent that they deal with given reality; moreover, we must lay down some rules which are to be ascribed to logic in the broader sense. Logic 'in the broader sense' encompasses not only formal rules of deduction, but also rules for the construction of an exact language. In particular, a closed and partially new account will be offered as a solution to the question of the logically and methodologically correct introduction of fundamental distinctions which must already be made for measurements to be possible. This methodology is presented in Section 4 of this chapter.

Since essential stimuli to this account go back to the works of Hugo Dingler and Paul Lorenzen and since there is a matter of some differences or extensions in respect to these preceding works, the theory of Dingler shall be presented in Section 2 of this chapter and its extension and development by Lorenzen in Section 3.

2. METHOD AS A VALIDITY CRITERION. ON THE FOUNDATIONAL THEORY OF HUGO DINGLER

2.1. *H. Dingler and Protophysics.*

In the first edition of this book, the section 'Method as a Validity Criterion: On the Foundational Theory of H. Dingler' led to misunderstandings. The most aggravating was the mistake of interpreting supporting passages in that section (substantially unchanged below) in such a way that the protophysics of time presented here merely became a realization of the older Dingler program;[12] with this some of the older arguments against Dingler would also supply arguments against protophysical chronometry. Here an elucidation is expedient which, after citing some examples for subsuming protophysics under Dingler's theory of science, emphasizes differences between the two while in turn also noting those aspects that have been taken over from Dingler.

First, then, we shall give examples where protophysics is classed with the Dinglerian philosophy, without additional differentiations. This occurs most wholesale, indeed, with W. Stegmüller, who wishes to have discovered a 'resurrected Dinglerism', which, in the form of 'constructivism', "exercises fascination ... on young minds' (Stegmüller, 1973, p. 25 f.). Though Stegmüller himself mentions no names and consequently practices a new style of academic controversy – after all he diagnoses a tendency towards 'dogmatism, intolerance' and 'sectarianism' (*Ibid.*), and it may remain open whether the condescension which lies in his claiming a psychological understanding of the young scientists with whose works he substantially does not come to grips is intentional – *The Protophysics of Time* assumes in many respects a central place among the writings that Stegmüller generally could have meant. However, as regards the chief reproaches of Stegmüller, we shall show below that there are decisive differences between this book and Dingler's works, which differences Stegmüller did not or would not see.

Besides numerous other authors who readily go back to Dingler when they think of protophysics – a reason for this may be that the criticism of Dingler already occurring in Carnap, Reichenbach, Popper and others may be cited – A. Kamlah, above all, objects to protophysics; its advocates appear to him simply as 'neo-Dinglerians' (Kamlah, 1975). Though he sees important differences between Dingler and the advocates of protophysics, he indeed treats the geometry as a 'Dinglerian-Lorenzenian' one. In Kamlah's later commentary on protophysics (Kamlah, 1979) in which work some older misconceptions no longer occur and in which in particular he no longer attempts to attack protophysics with physical arguments, Dingler's name hardly occurs anymore; indeed, in the appendix Kamlah expressly points out the danger of overlooking the difference between Dingler and protophysics. Nevertheless his objections still touch on claims advanced by Dingler and not by the advocates of protophysics. This holds in particular for the assertion that protophysics is the sole possible operative foundation of physics, although this has been asserted nowhere by any of the authors cited (Kambartel, Lorenzen, Mittelstrass, Janich). Kamlah constantly confuses, on one hand, the assertion of the *impossibility* of a system of norms rival to protophysics (he thinks thereby of "a system of norms for special relativistic physics", "whose feasability ... stands in contradiction" to the feasability of protophysical norms) with the assertion that no rival system exists at present.

This confusion rests on the fact that he does not see the differences in goals: He seeks a new norm system in the context of an affirmative fundamental approach (cf. Chapter I); I seek one with a philosophically critical intention that cannot accept as an unexamined presupposition the complete reconstructability of presently existing physics.

Consequently it is already a matter of setting out in the following differences between Dingler's position and protophysics. Since the relationship of Dinglerian philosophy and the constructivism of the 'Erlangen School' has already been explained by J. Mittelstrass, and since he has demonstrated there that whoever carried over a criticism of Dingler to constructivism "(has) not understood the intentions of a constructive theory of science or ... consciously (reinterprets) it for polemical purposes" (Mittelstrass, 1975, p. 25 f.), we need merely enumerate some points here, which we will then expand only for those that especially concern protophysics.

Differences between the Dinglerian philosophy of physics and protophysics to be held to are:

(1) The start of methodological construction in protophysics is not marked out voluntaristically or decisionistically. In protophysics it is not at all a question of giving a 'final foundation' in the Dinglerian sense, but rather of answering the question of how a measuring physics would be possible. This question itself presupposes, among other things, that criteria for the intersubjectivity of a measuring physics have already been established, as well as a life-world (lebensweltliche) basis of acting and speaking with each other. Marking out an 'absolute basis' is not the business of protophysics. It rather understands itself as an endeavor that is auxiliary to physics, where physics for its part furnishes technical knowledge and where one is never to lose sight of the fact that technical knowledge is not the whole of knowledge.

(2) Dingler appears in the course of his scientific life to have come to the opinion that his way was the only possible way to ground physics. It can remain open here whether Dingler's critics always have come to grips earnestly enough with this claim, where broad agreement for a superficially based rejection of it could already be reckoned on. (Cf. Diederich, 1974, without presenting an argument for it, in a contrast of Dingler's views with classically conventional ones, in the sense of Poincaré, D. writes that Dingler's philosophy of science is "materially less interesting" (p. 134). This estimation is due to an attitude that expresses itself in sentences like "Insofar as both methodological standpoints (sc. Popper's and Dingler's) are normatively intended ... it naturally cannot be objectively decided between them and

they shall not interest us further here". (p. 143). At the stage of the *present* state of discussion relative to normative methodologies, satisfying oneself with an appeal to an old positivist consensus (normative = naturally not objectively decidable), can scarcely avoid giving the impression of superficiality.) Never yet, however, has any advocate of protophysics advanced it with a claim of unique validity. As with every theory presented in argument, it is subject to the danger of human error; as with every theory it is subject to the revision of old insights by newer ones.

Indeed it can already be maintained here – hardly different than in Dingler's time – that up to now there are no alternative proposals that compete with protophysics and that have comparable aims. When e.g., A. Kamlah asks: "Can Janich prove that no alternative is open, that there is no other system of norms" (Kamlah, 1979, p. 322) than that of protophysics, this is an academic, inconsequential question. The claim imputed here has not been advanced, nor has, up to now, a competing system of norms, e.g., one that leads to relativity theory, been given.

(3) Dingler argued against Einstein's relativity theory and later certainly also polemicized against it. Since we are concerned here with the relationship of Dingler to protophysics, we shall not be interested in whether polemic is intelligible or excusable if existing arguments do not seem to be considered nor shall we be interested in whether Dingler's polemic has been splendidly corroborated in the retreat of the empirical-analytic theory of science of Russell, Schlick, Reichenbach and Carnap to Kuhn, Lakatos, Feyerabend and Sneed. What it comes down to here is the fact that protophysics – in spite of commentaries to the contrary (Thus, e.g., G. Böhme asserts that protophysics burdens itself with a polemic against the theories of relativity.) – is not burdened with the polemic against relativity theory. Critical remarks on relativity theory made by advocates of protophysics have always been concerned with the art and manner of how, in an empirical theory, one has attempted to do justice to the historical fact of relativity theories.

(4) Protophysics always has striven only for operative foundations of physics and has granted these a special status over against empirical physics i.e., physics based on measurements. It has, however, never asserted the *a priori* character of all physical theorems. To be sure it is not possible, without lengthy additional explanations, likewise to characterize the status of protophysical norms as *a priori*, – along with the theorems e.g., of Euclidean geometry, according to the Dinglerian view; but in any case, in Dingler, theorems of Euclidean geometry and the Newtonian law of gravity have the same status; in protophysics, on the contrary, they do not. In Dingler

non-Euclidean geometries can acquire no empirical relevance; on the contrary, in a physics based upon protophysics they can do so.

It may be that, even when objections directed against Dingler are no longer raised against protophysics, there still will remain objections to protophysics. Yet it would certainly be a gain if protophysics were criticized on account of its own assertions and proposals and not on account of those that it does not advocate. Metaphorically speaking, protophysics stands upon the right of sons not to be measured by judgment passed upon their fathers, where the sons certainly best know the differences between their fathers and themselves. But, staying with the metaphor, grown sons must not deny the virtue of their fathers when the sons go their own ways.

Now as before, there is no cause to abandon the crucial insights of Dingler. These also appear again in protophysics:

(1) Physics, which we encounter as a specific result of its history, must be methodologically reconstructed or understood if it is not to be assumed (merely historically) as a simple fact. Though Dingler did not formulate this with such sharpness, one may distinguish between philosophical efforts to understand science as a result of its past, and, on the other hand, efforts to cultivate science with the help of its claim. Dingler, as do not advocates of protophysics, attempts the second. If, in contrast to the advocates of protophysics, Dingler also may have seen an alternative between the two activities, it nevertheless remains his merit to have recognized for the cultivation of science, the precept of methodological order in its role which is constitutive for the claim of the exact sciences.

(2) With this he also stated the decisive reason for an anti-empiricist attitude: experiences with any sort of claim to be called exact already precisely presuppose the formulation of this claim. This argument alone suffices to avoid empiricist (and, what Dingler no longer actually considered, critical-rationalist) paths. Protophysics like Dingler is bound to the effort of clarifying the presuppositions that must be made by man for an empirical-scientific acquisition of knowledge.

(3) Finally, Dingler, just as the advocates of protophysics, recognized the role of an operational foundation of physics. The objects which physics speaks of are 'natural' objects like the sun or moon only in exceptional cases. As a rule they are produced for experimental arrangements by manufacturing efforts. To be sure, the history of philosophy of science is itself burden by a kind of operationalism that has become irrelevant as a consequence of some exaggerations, namely by W. Bridgman's position who, by allowing arbitrary 'paper-and-pencil-operations' has stripped operationalism of its potential for

the foundations of experimental science. That physics owes the constitution of its objects to 'operations', namely manufacturing actions (*herstellenden Handlungen*), was seen by Dingler and is again incorporated in protophysics.

After this introduction to the excursus below on Dingler's answer to the question of the presuppositions of time measurement – which excursus was written almost 20 years ago now – perhaps the following can be read under changed presuppositions. This introduction was also conceived as an answer to some objections to protophysics.

2.2. *H. Dingler's Foundational Theory*

Dingler's many-faceted work, which is concerned primarily with critical foundational studies and consequently extends from logic via mathematics to the empirical sciences, encompasses a large number of publications special to foundational problems relevant for physics. This work stretched over a period reaching from 1907 to Dingler's death in 1954. (An introduction to the methodological philosophy of H. Dingler is found in the introduction by K. Lorenz and J. Mittelstrass to H. Dingler, *Die Ergreifung des Wirklichen*, Chapter I–IV, in *Reihe Suhrkamp Theorie I*, Frankfurt 1969, Dingler's most important publications are also listed there.) It only found its conclusion with the publication of *Aufbau der exakten Fundamentalwissenschaft* (1964) by Lorenzen. It may be agreed clearly that the basic conception of a theory of the foundation of the exact sciences, which Dingler may already have developed in 1905, has hardly changed over this long period; however, in questions of detail, a developmental process did occur.

Nevertheless we shall not write a biography of Dingler here and shall leave historical details unconsidered (in this respect we find ourselves completely in agreement with Dingler, who always intended science to be independent from historical contingencies). Thus we will place as little value on a chronological presentation of his work as on a discussion of contingent developments. Meanwhile the attempt to present his point of view as a closed whole and to omit, from the outset, proposals of solutions which the author himself later reconsidered promises to be profitable.

The investigation of Dingler's theory of science will be restricted to a domain more closely defined as follows: From the outset, a division of theoretical discussions which methodologically precede the establishment of empirical laws can be made by distinguishing physical concept formation from other preliminary judgments. This distinction essentially underlies

Dingler's *Physik und Hypothese*. (1921, in the following cited as PuH). It implies the following: What one ordinarily designates as physical 'laws' are, logically considered, universal propositions. (This also holds for the more special term 'general proposition' in *Die Grundlagen der Physik* (Berlin 1923), which term signifies independence from place and time.) In order to be able to obtain empirical universal propositions, (1) reproducible and therefore comparable measurements must be possible, (2) the conditions of experimentation must be elucidated (for example, the principle which Dingler called the 'identity theorem', i.e., under similar circumstances similar effects occur) and (3) the role of hypotheses for establishing propositional systems must become transparent.

What has been programmatically conceived of as 'protophysics' by Lorenzen (Cf. Chapter II, p. 3.) merely encompasses the first point, namely physical concept formation. With protophysics, it is a question of completely providing the presuppositions of reproducible measurements. Since this book likewise devotes itself solely to this task, *Dingler's conception will be investigated only to the extent that it covers the possibility of comparable measurements*. With this is stipulated that Dingler's foundations theory will be pursued only to a level where 'objective' individual judgments can be seen as possible. To be sure Dingler himself did not draw this boundary; nevetheless the question of the possibility of 'objective' measurements may be answered in itself without falsifying his conception. We will prove in the following account that no physics of whatever nature can be established without clarifying how comparable measurements are generally possible. That, beyond this, further questions must be clarified for the path to empirical universal propositions to be pointed out step-by-step, however, no longer will be the subject of this investigation. (Dingler expresses himself on this in e.g., PuH, p. 42–111.)

In spite of objections yet to be raised against the construction of Dingler's methodology, it is fully compatible with his work to first present the methodology in general, to next consider more closely and individually the relevant points of view – particularly those relevant to the foundation of physics – and then to finally examine to what degree the use of method in constructing physics prospered with him.

In overview the following picture presents itself: The historical rise of physics is dependent, in its development, on various accidents; it encompasses false starts as well as leaps of genius which were only justified later (or have still to be justified). In contrast to a historical investigation of the realization of an intellectual structure such as that of physics – a structure which is justly a model for all other empirical sciences – the aim of Dingler's research consists "in the intention of showing that it is possible to establish an enduring, so to speak absolute, science apart from and logically independent of all sciences which have become historical and conditioned by cultural contingencies". (*Grundlagen der Physik*, 1919 and 1923, p. V, in the following cited as GdP). (If one attempts to restrict his contemporaries' judgment of Dingler's theory to scientific or philosophical questions, i.e., attempts to disregard Dingler's role in the dispute over 'German physics' during the Third Reich, then this claim – namely of constructing a final science which is not compelled to make revisions by any progress in empirical research – has, for the large part, subjected Dingler to criticism, above all from physicists. Indeed the physicist's position proceeds from the empiristic misunderstanding that there could be empirical scientific knowledge even without preliminary decisions which are not revisable *by measuring experience*.

Dingler tries to do justice to this claim of certainty by questioning all the steps in the construction of the exact sciences and by again investigating all foundations in turn in respect to their own foundations. He shows, finally, how one escapes the dilemma of having still to provide a foundation for the initial elements themselves; the dilemma arises because there cannot be infinite foundational chains. Dingler calls the goal striven for 'full foundation' (*Vollbegründung*) and achieves it by means of the 'synthetic method'; this method advances to secured results by building upon a basis in steps (e.g., logical rules), which for their part are traced back to the initial basis. As for what in particular concerns physics, Dingler hopes to answer the question of how theories qua propositional systems can be applied to reality; citing Kant, he himself designates this as the task of "submitting to basic principles the transition from the rational to the empirical". (GdP, p. 36).

It is easily seen that one of the chief difficulties will lie in pointing out what may be taken as the beginning of the synthetic construction and why. Here is an appropriate starting point for a detailed presentation of Dingler's methodology.

Independently of every epistemology or, in general, of every philosophical doctrine that always asserts propositions as true – which immediately raises the question of their foundation – man finds himself on a naive level of thinking; as long as he does not speculate about thinking itself, as long as he claims no knowledge of knowledge itself, in short as long as he does not reflect, he remains on this level. Dingler calls this level the 'life standpoint' (*Lebensstandpunkt*) (PuH, p. 114.) Or 'standpoint of daily life'. (*Aufbau der exakten Fundamentalissenschaft*, 1964, p. 42; in the following cited as

FWS.) "Everyone of us is in this standpoint almost constantly, as long as one does not reflect on this "life" itself". This is not to exclude the fact that a man in the life standpoint thinks or reasons; only certain, namely 'self-referential', thoughts and reasonings are excluded.

In spite of a definition which sounds different, the life standpoint coincides somewhat with the 'pre-general standpoint' (*Vorallgemeinstandpunkt*), which is the starting point for the foundation of physics in Dingler's other presentations. Whoever does not regard any sort of universal proposition as in any way secured, whose reasoning concerns only individual objects and who consequently knows no facts other than singular facts, finds himself taking this standpoint. (Cf. GdP, p. 8, 12.)

It is common to both standpoints that in them man has at his disposal practical abilities which enable the functioning of the life process (*Lebensablauf*) and which are independent from every theoretical foundation. Only where unreflective living is possible can a person go on to cultivate science.

To these abilities belong all types of action, in particular also those which must be carried out in order to construct an exact science of reality, i.e., linguistic actions and manual operations (for producing measuring devices). Here it will already be evident that the prescientific abilities cannot be established theoretically as 'earlier', since how else is this supposed to happen than in linguistic actions.

The ability to have (prescientific) singular experiences and to mark them also belongs to the 'basic abilities' (FWS, p. 35.) (*Grundfähigkeiten*) of man. The question, then, immediately suggests itself whether or not singular experiences or a greater number of such could yield a solid basis for general propositions. The view readily advanced in physics textbooks that an 'overwhelming abundance' of experimental data (singular experiences) finally allows no other choice but to consider a law (universal proposition) as valid is a conception of this type. This path is indeed strictly not logically possible, and since Dingler does not ask for more or less well confirmed inductively acquired propositions, but asks for valid universal propositions in the strictest logical sense, a probabalistic applied inductive logic, say in accord with Carnap's proposals, does not come into consideration for him. Dingler repeatedly opposes empirical attempts at foundation, whether undertaken, on one hand, in positivistic empiricism in the form of a sensualism or a psychologism, or, on the other hand, indirectly via 'ontological' doctrines.

The alternative, namely to renounce every foundation of initial universal propositions and to assert them 'axiomatically', is likewise unacceptable, since then every claim to certainty would have to be given up. Axiomatic theories are always only as certain as their 'weakest point', which point is to be sought in the unfounded axioms themselves. If the axioms remain unproved, then every theorem derived from them is also unproved; the theory is merely a 'hypothetical formalism'.

Resigning oneself to the renunciation of universal propositions would finally imply the renunciation of every science and, beyond that, would generally prohibit every action of securing knowledge. The logical laws which are drawn upon in the inferences for the foundation of even the most simple theorems are already general propositions.

One may assert that Dingler succeeds so well in arguing against the most workable doctrines of how valid universal propositions are possible that an honest adherent to such doctrines, especially the empiricist, is led into aporia. It is certainly easy to dismiss Dingler's presentations as an exaggerated criticism so long as no way out of the dilemma of missing foundations, above all of physics, offers itself. However Dingler always saw his problem as the construction of a fully founded science. The (occasionally polemical) attacks on other, and as has been shown, unworkable attempts at foundation therefore certainly must be seen in historical context for them to be able to be judged objectively. However, as was established at the outset, this task shall not concern us here. Dingler's disputes with other doctrines run parallel to his own construction; if these disputes are not mentioned, his own construction will contain no holes.

The way out of the foundational dilemma of the exact sciences indicated by Dingler consists in a 'critical voluntarism'. (GdP, p. 12.) First universal propositions are 'supposed' in a free decision of the will. The central role of the free will of a person striving for science – from this Dingler obtains the beginning to every exact science – requires a deeper penetration into this part of Dingler's theory; for a reason yet to be presented, this part is not without its own problems.

No concept belonging to a psychological or philosophical theory is meant by 'will'; rather it is a question of the spontaneous ability with which man 'naively' serves himself in the life standpoint. (GdP, p. 11.) Thus it counts among the basic abilities and in that respect is not to be questioned theoretically in respect to what stands behind it. Every attempt to show a dependence of the will on any presuppositions whatever, say on singular experiences, must – so Dingler's language is to be understood – itself again return to a spontaneous decision. Likewise, objections of the sort that the will is not free may not be advanced, since with these too it is again a question of a universal proposition requiring a foundation.

The whole discussion revolving around the 'will' serves finally the purpose of demonstrating the dependence of every science on human actions (Handeln); this must be seen as an alternative to the methodologically untenable conception of 'the world' obtruding its lawfulness on the passive observer, of its presenting itself as controlled by laws. (Cf. FWS, pp. 20–24.) "Reaching valid propositions in general" succeeds by means of a "resolve of the will". (GdP, p. 11.) Here and, above all, later in FWS – where Dingler distinguishes between what is subject to our will and consequently available with unassailable certainty and what our will does not achieve – the following difficulties appear:

In striving to comprehend, in a methodological discussion, the conditions for the possibility of universal propositions, he himself speaks in generalizing form of 'our will'. Instead of directly demonstrating the activity which sets science in motion, he speaks about this activity and, with the sincere attempt to prove beforehand all subsequent steps as conclusive, creates for himself the difficulty of being able to be interrogated about the foundation of his methodology. Even though the essential progress of Dingler's philosophy lies in the fact that he assigns to unquestioned actions a place alongside mere linguistic fixations of states of affairs, he refrains from using his own achievement in his methodological discussions – something which leads to some inconsistencies.

This critique, let it be noted, applies to the construction of his methodology – one could almost say to the arrangement of his books. (A remark indicates that Dingler himself already saw the objections raised here. After presenting what role human activity plays in the construction of exact science he admits: "Thereby we already must make use of some rough outlines of that which we wish to attain (namely of fully founded science). Thus the presentations containing such outlines bear only a suggestive character here, in order that they will be intelligible to the reader. The actual taking of the first step of the construction occurs in the next section. There also then the will makes its appearance just in its purely active form". (FWS, p. 27)). The method itself, however, is attractively unassailable; for the beginning on which exact theories are constructed is unassailable (in the sense that one cannot go behind them theoretically), and the individual steps which are taken according to the principle of pragmatic order are unassailable.

'The beginning', i.e., the universal propositions volitionally guaranteed in their validity are namely nothing other than terminological regulations (Festsetzungen); thus they make no claims to truth. Dingler gives the example "a rectilinear figure with three corners I call a triangle". Now since the division undertaken from the ntandpoint of daily life – i.e., '(a) into that which is subject to our own will and (b) into that which is not or not completely subject to it" (FWS, p. 22.) – is regarded as a complete division by Dingler (i.e., there is nothing which does not belong to one of the domains (a) or (b)), terminological definitions are the only propositions which can be secured 'prelogically'; moreover, due to their independence from further foundations, their validity (Geltung) is to be called 'absolute'. (FEW, p. 22, 23.)

Here, it may be said, the already mentioned weakness in Dingler's presentation of his method – which method in itself is non-circular – completely shows itself: The reader of his works is in fact justified in asking how the statement that everything is either not or partly or completely subject to our will is to be understood. This is a case of a universal proposition upon whose validity all connected conclusions depend, in particular the assertion that his method is the only one possible. To prove this validity Dingler must terminologically define 'will', present a theory of the will and finally analyze attributions such as "that which is subject to our will . . .". He must, in short, found his theory by means of a new theory.

Nevertheless it must be emphatically pointed out that only the presentation of the method, i.e., the verbal description of the methodology, must be regarded as a failure; however, with the restrictions stated on pages 30 to 35, the method itself does justice to the claims with which Dingler presented it. An example may illustrate the problem of Dinglerian methodology: Whoever writes a methodology which consists in nothing other than a discussion of the order of individual steps yielding chains of arguments and in the furnishing of the first step, finds himself in a position similar to someone who composes a book entitled *How to Read*. As little as a 'reader' could learn to read from this book, it could nevertheless be meaningful – say, within the scope of a psychological investigation for pedagogical purposes. Indeed it is only intelligible to someone who can already read. In the same way every linguistic representation can be understood in its scientific exactness only when elementary rules of a methodology have already been accepted, in particular a metarule for the uniform use of words. Such a rule can only be learned naively and unconsciously with the acquisition of language; it cannot be learned indirectly via linguistic instructions, rather only though demonstration and imitation and only subsequently can it be formulated with the already available linguistic means.

Let it be noted here that Dingler is not burdened with the unresolvable dilemma of presenting a logical propaedeutic, as the art of rational speech, that strictly is not able to be taught in a book, because it would rely already then on the reader's at least partially mastering it. In contrast, a strict methodology, something which especially concerns the method of physics, is very well possible *before* beginning physics (as an alternative to a metalinguistic account after the fact, or to a paralinguistic elucidation which parallels the construction of physical theory). To be sure, for this a developed logic in the broader sense is needed, a logic which has already solved the problem of making valid general propositions possible. Since, nevertheless, the elements for the construction of an exact science have already been established (namely terminological definitions), the discussion of the order of the subsequent steps can commence.

Dingler formulates a 'principle of pragmatic order' (Cf. Dingler, *Die Methode der Physik*, Munich 1938 (in the following cited as MdP), p. 116, 117 as well as FWS, p. 26.) which states that every step may only be constructed upon steps which have already been carried out. This is equivalent to the prohibition of pragmatic circles. Here 'pragmatic circle' is a general concept for definitional, logical and practical circles.

Practical circles, for example producing a first ruler with a machine whose own production is impossible without a ruler, can, as it were, be run through only in false propositions, i.e., only on the linguistic level; they cannot, however, be run through in fact. It is from just this true, because tautological, proposition that the functioning of physics and technology may be understood in spite of some false methodological presuppositions.

It is a tautological assertion that adhering to the principle of pragmatic order can lead only to fully founded theories. The criterion for fully founding theories may be expressed conversely by the methodological requirements that (1) running through a chain of arguments in a backwards direction (say in a dialogic situation by means of consistent questioning) must reach an end, and (2) that this 'end' may itself not be an assertion which claims any sort of truth, i.e., which would require criteria for its validity.

Consequently the synthesis of a fully-founded theory is, indeed, at the outset only tenable as far as the construction of logic and mathematics, i.e., formal sciences, reach. The central problem of empirical science, namely how the newly constructed language becomes useful for exact propositions about the given (*Vorfindliches*), a problem also understood as the question of the 'suitability' of a representation, has not yet been touched. It was Dingler's merit to have pointed out a way to solve this problem, which problem he called the 'application problem'; (GdP, p. 35.) This solution again rests on the fact that manual creation, manipulating the enivornment, occurs alongside linguistic action.

A merely intellectual orientation in the non-scientific world did not lead historically to physics; nor can a speculative – i.e., passively contemplating and describing – activity methodologically give rise to physics. Rather, active intervention is also a precondition; i.e., changes in the world of bodies and processes, in particular the production of measuring devices and the creation of experimental conditions are also preconditions.

Thus it is not that given 'real things' are indiscriminately predicated, rather, after establishing general propositions through convention, those real things are selected "which obey these general propositions. We designate this procedure as the principle of realization through selection". (GdP, p. 37.)

To begin with, we have not gained much with this, because the problem consists precisely in how we can decide whether real things 'obey' general propositions. Dingler's works do not offer a unified picture on this point. First the term 'realization' is used in the domain of actual physical concept formation as well as in the domain of mathematical concept formation. Thus, for example, the operation of placing marks one after the other and thereby producing numbers is also called 'realization'. (See FWS, p. 62ff.). Yet even restricting ourselves to physical basic concepts yields no general use of the word. Indeed, in the domain of geometry Dingler constantly mentions the three-disk process[13] for producing planes, a process in which steel disks are ground against one another until they fit on top of each other. Also Dingler's remarks on chronometry in the end amount to the construction of real clocks; nevertheless numerous passages may be produced where 'realization' signifies merely the mental fulfillment of a construction plan of an exact theory. Thus the realization of the 'simplest' motion is uniform motion, which, now, is not to be understood as say the motion of a clock's hand or of any sort of real body, but is to be understood merely as the imagined motion of a point an imagined line. (FWS, p. 188, 189.)

In a discussion of the problem of application, as one may expect, 'realization' only can be relevant as the procedure of working on real bodies or influencing real processes. Now under this restriction, Dingler understands by 'realization' not only the procedure of producing a realization once and for all – for example, planar disks –, but also the whole refinement process carried out with the advance of experimental technique, which process necessarily leads to ever higher accuracy in measurement and consequently to every finer realizations.

First, as for the realization ***procedure*** – e.g., the one-time production of planar disks, a clock or suchlike –, it ***uniquely*** determines the form to be realized. (MdP, p. 109.) To be sure we cannot exactly say how Dingler uses the term 'unique'. In the given context, the uniqueness and definiteness

(*Eindeutigkeit*) of the realization procedure, which uniqueness depends on the unique determination of the forms to be realized, could mean two things: Either it means that the given definitions suffice, i.e., there cannot be – to choose a geometrical example – different forms of surface boundaries of bodies which satisfy the definitions of, say, the plane – the indistinguishability of the points of a surface is, then, not a unique definition in this sense since it holds for planes and spheres –; or 'uniqueness' means that carrying out the realization procedure leads to realizations whose relations to each other are established *a priori*. No evidence whatever is to be found for this second version, which is an essential component of the methodology presented in this book; thus one must interpret Dingler accordingly as saying that the uniqueness of the realization procedure, as a consequence of the unique definition of the forms to be realized, signifies the completeness of the introduction: 'uniquely defined' would then be synonymous with 'fully defined'; 'uniqueness' would be nothing other than 'full foundedness'. (To be sure in FWS it is proved that "one and only one plane is determined" by three points which do not lie on a straight line (p. 182); yet here it can only be a question of a geometric theorem which is formulable and provable only after introduction of the basic concepts plane, straight line and point. Nevertheless, already for methodological reasons, this proof does not show the uniqueness of the definition of 'plane', because this uniqueness has already been presupposed for the definition of a straight line as the line of intersection of two planes.) Above all, the use of 'unique' in FWS supports this interpretation; there also the terminological definition of 'body' and the various possibilities of distinguishing boundaries of bodies are designated as 'unique'. (The definitions tied to the presystematic language of daily life "do not state something about the nature (*Beschaffenheit*) of reality; rather among those uncounted possibilities they select only definite ones, in order to define uniquely by steps specific concepts in any respect". (FWS, p. 175.)

No further discussion is needed, then, say to prove uniqueness understood thusly. If all terminological definitions are traced back to distinctions available in the standpoint of daily life, uniqueness is secured en passant. It may already be noted here that the assertion of the unique reproducibility of the basic forms which enable measurement needs a certainty which exceeds what Dingler ascribed to it.

Thus if one holds to the proposition that, in Dingler's account, proofs for the fact that repetitions of a realization procedure yield the same results do not appear necessary, because the forms to be realized are fully defined, then the impression is strengthened that Dingler was not able to come round to a clear decision about what significance should be attributed to the realization procedure now understood as the manipulation of real bodies, i.e., how important the discussion of the realization procedure is for the methodological construction of physics. In spite of emphasizing how significant active intervention in the environment is when it is a question of the foundation of physics, one must nevertheless understand 'realization' as a 'mental' process. Otherwise Dingler's statement cannot be explained when he writes in *Foundations of Geometry* (Stuttgart, 1933): "Our definition of the plane is independent of the triple disk procedure described". (p. 42)

Knowledge of the three-disk procedure for realizing a plane is then just a happy historical accident, especially as it represents "exactly the realization of our . . . definition of planes". (GdP, p. 38.)

Presumably Dingler would have still considered a science of reality constructed upon indistinguishability requirements as a meaningful undertaking, even if there were now no procedure which could be directly transformed into manual activities, but only one which could be carried out mentally and thus would amount to a logical connection of ideas in the Dinglerian sense. ('Ideas', which are also designated as 'forms' (FEW, p. 4), take over the task of goals (Zielvorstellungen) for realization procedures. (FWS, p. 46.) Thus in his later writings Dingler more ignored than solved the 'application problem', which problem he designated as significant in his earlier works. Thereby one still stands in agreement with Dingler if one sees the 'application problem' as consisting precisely in attaining certainty for talk of the world appearing in the life standpoint – world with more or less hard bodies and the most varied events – by actively intervening in the world of 'real things', i.e., by grinding disks on one another and forcing motion forms on bodies.

Thus if Dingler did not state with necessary clarity what realization procedures ought to be, the place they assume in the construction of exact theory is nevertheless clearly comprehended. In the realization procedure there must be "contained also all the necessary elements from which the properties of the realized forms logically follow". (MdP, p. 109.) This means, for example in the case of geometry, that a suitable representation of the realization procedure for the basic forms of geometry implies all the theorems of (Euclidean) geometry, in other words the propositions describing the realization procedure "must represent an axiom group which yields geometry".

Though Dingler considered it an essential component of a strictly carried out methodological construction of physics to explain and to incorporate in his system the successive refinement of the realization procedure as methodologically possible and as proceeding in an assignable direction – for this reason he laid down a 'principle of levels of accuracy' and explained the improvement of realizations by means of 'exhaustion' of established theories – this part of Dingler's methodology shall be presented only in overview; as mentioned at the outset, the attention within the scope of this book focusses primarily on the area of Dingler's philosophy that is tackled anew and developed here. Discussion of his methodology is continued in the respect that we consider its application in the construction of an exact physics, a construction undertaken by Dingler himself. In this construction a leading role falls to geometry, since all measurements finally amount to comparisons with rigid bodies.

Ignoring some details, Dingler's work on geometry may be summed up as follows: In the 'pregeneral standpoint' we are acquainted with bodies of various hardnesses. (For Dingler even the concept of body is an 'ideoid' and he presents an extended definition (FWS, pp. 156–175) in which talk of bodies and their (surface) boundaries is reduced to the basic concept of distinguishability (of e.g., solidity). Such an exact definition may be required, if, for example, it is to be decided whether elementary particles are bodies; however, in my opinion, for the construction of geometry it is not required.) By 'hardness' is meant a practical rigidity, i.e., a property of existing bodies. If, now, in the domain of practically rigid bodies we are to seek the most rigid, it then becomes a question of searching for the best possible realization of an ideally rigid body; thus, trivially, a definition of 'rigid body' must be available, otherwise a meaningful search for a realization is impossible. (The complete construction, which begins with given 'solid' bodies, leads via a selection norm (cf. p. 146 of this book) to the restriction to 'hard' (relatively to each other, of constant form) bodies, and finally to the definition of 'rigid' (constant length under transport) bodies, is found in Janich, 1976a.)

As a definition of rigidity Dingler selects the requirement that a body must obey the laws of Euclidean geometry. In accordance with the principle of realization through selection, a real body is shown to be rigid in its agreement with geometrical theorems. In practice this procedure occurs in individual steps, the taking of each of which signifies respectively, a higher precision of the realization. Thus a 'hierarchy of rigid bodies' may be produced, the momentarily most rigid of which respectively determines the highest attainable accuracy of measurement. This increase in the values of realizations through refinement of the control procedure on the basis of already realized forms, and the bounding of the respectively highest measurement accuracy, is an infinitely continuable process. Dingler considered it as actualized in the practice of producing measurement devices and interprets it as if it were guided by a 'principle of levels of accuracy'.

If now, in the real case, with the control procedure there are ascertained deviations, they must be traced back to disturbing circumstances. Such explanations of deviations by means of disturbing circumstances fulfill the 'principle of exhaustion'.

Though Dingler himself does not make the following distinction and though it must remain open whether he intended it, a sharp distinction of classes of situations in which exhaustion must occur is practically possible: on one hand with the realization of basic forms such as planes or uniform motions, on the other with the experimental testing of hypotheses, say of the Newtonian law of gravity. (This distinction, which is crucial for judging Dingler's philosophy and upon which it finally depends whether or not there can be empirical physical laws at all, is overlooked by some critics of Dingler, thus e.g., in the review of Dingler in the otherwise very worthwhile book of Diederich, 1974.) In view of our intentional restriction to physical concept formation, omitting further discussion of how physical propositions are established, here we will only consider exhaustion in the realization procedure of basic forms.

Dingler defines: "By exhaustion we understand the procedure of introducing into actuality fundamental concepts of the unique system and their consequences, i.e., such concepts which are primary within the system, in contrast to all further concepts. This introduction occurs mentally, in that the concepts and processes, as far as they are relevant, are brought into account and everything that is not yet covered by them is defined as a superimposed phenomenon". (MdP, p. 146.) Dingler demonstrated this procedure in detail for the concept of rigid body in PuH and in GdP. With the production of a rigid body through exhaustion of Euclidean geometry, Euclidean geometry is simultaneously "introduced into reality". (PuH, pp. 26–29.) And: "The logical concept of a rigid body is fully equivalent to that of a 'geometry' ". (GdP, p. 127.) However, how one obtains the theorems of a geometry which one certainly already needs to introduce the term 'rigid body' is already indicated: They are derivable from a description of the realization procedure of the geometrical basic forms. Such descriptions are formulated as indistinguishability requirements.

With this view Dingler placed himself in clear opposition to the then as now prevailing opinion of the 'structure of physical space'. Consequently he did not neglect proving the impossibility of a methodically ordered, and hence intelligible, construction of a unique geometry on the basis of pure empiricism. We need not repeat here the proof of why it is impossible, purely empirically, to establish a strict decision procedure for deciding which of two bodies is the more rigid, since as it stands, this claim is no longer raised in present-day physics. There one believes oneself freed from this problem through a geometry based on the phenomena of light dispersion, indeed forgetting thereby that an optical geometry cannot be obtained other than via a geometric optics, which for its part cannot manage without rigid bodies (for diaphragms and screens, for interference experiments and measurements of light velocity).

The alternative procedure to the principle of exhaustion – not searching for disturbing circumstances when real bodies deviate from the laws of Euclidean geometry, but rather changing the geometry – according to Dingler, can never lead to unique theorems. (Cf. Dingler's objections to the view of H. v. Helmholtz, MdP p. 202 ff.). Completely apart from the fact that measurement devices always are produced de facto according to the principle of exhaustion of Euclidean geometry (other measuring devices could, on trivial grounds, produce no comparable measurement results), no universal proposition can be formulated free from geometry, which proposition states from what deviation on the geometry is to be changed.

The task still remaining in the construction of geometry now having been outlined, namely deriving an axiom system from a description of the realization procedure, was tackled by Dingler himself. He took Hilbert's 'axiom of connectivity' to be proven through his terminological definitions of planes, lines, etc. It would be a certainly worthwhile task to investigate to what extent Dingler's works on this point (in FWS, pp. 160–172 and pp. 175–187) are really stringent and to what extent they represent perhaps a basis for further promising work on a methodical construction of geometry. In regards to the theme of this book, however, let the above remarks on geometry suffice.

As for what concerns Dingler's foundation of a theory of time or of chronometry, his investigations have remained fragmentary. Since time-measurement 'locally' needs ample elucidation before problems such as 'simultaneity', 'clock transport', 'time comparison at different places' among others can be taken up, let the discussion of Dingler's theory of Chronometry remain restricted to this point.

That there is motion is one of the individual experiences in the standpoint of daily life. "Motion, in itself, is an elementary experience, i.e., it needs not and cannot be constructed and derived from other elements". (MdP, p. 113.) To be able to speak of 'uniquely determined motion', a 'new element', time, must be introduced. A 'presystematic concept of time', 'primary time', is constituted by enumerating 'full acts of will', ordered according to earlier or later. (All citations from MdP, pp. 113–119.)

"If now it is attempted to introduce a measure of time, there is no other means than the fact that one finds processes which constantly repeat themselves in the same manner, so-called periodic processes". The 'basic principle' of 'artificially constructing and isolating' periodic processes is based on the following idea: "If two processes which have a natural instantaneous beginning and a natural instantaneous end are so constituted that all their essential conditions are strictly identical, then the beginning and end of both enclose the same time interval". This 'principle of the equal duration of idential processes' produces a "definition of equal time intervals, whether both processes occur simultaneously or anytime after one another, whether they occur at the same or in different places".

He continues: "The definition obviously has the property of constantly increasing in accuracy with the progressive construction of unique physics – a process which, due to the unboundedness of physics, is itself unbounded. In that namely new circumstances are always discovered which can influence the course of the process and that mastery of these circumstances is attained, the identity of the essential (i.e., those influencing the process within the accuracy of the moment) circumstances can be better guaranteed". (This definition has, among other things, the defect that the identity of all essential conditions is required, but a criterion for essential conditions is not given. Thus, in this place, one simply has to take note that Dingler did not count difference of place among the essential conditions.)

After equal durations are defined in MdP via periodic processes, and 'time' is introduced systematically thereby into the construction of physics – which then allows motions in their course to be comprehended uniquely (MdP, p. 126) –, this conception is extended (Dingler, *Die Ergreifung des Wirklichen*, Munich, 1952.) By adding that only a successive refinement of chronometry by using clocks of every higher frequency, up to the degree where points of time are 'everywhere dense' within the periods of a slow clock, defines 'ideal time'. (*Loc. cit.* p. 73.) With this concept, then, uniform motion could be defined. Böhme correctly points out that this definition is nonsensical because between 'densely lying' time points there can lie *per definitionem* no duration that is determined by the periods of a motion, (G. Böhme, 'Protophysik der Zeit, eine nichtempirische Theorie der Zeitmessung?', *Philosoph. Rundschau 20*, 1973, 94–111, p. 102; again in Böhme, 1976, p. 287).

Dingler's conception of the presuppositions of time measurement, outlined in citations from MdP in particular, is not presented in his other works in ways differing in essential points from that in MdP, so that we can refrain from reviewing them. Independently of the adjacent discussions of ideal simultaneity and the objections to the relativistic idea of time, we must repeat doubts about Dingler' statements on the topic of time.

Following the principles of Dingler's methodology, the impossibility of non-circularly 'constructing and isolating' periodic processes must be evident immediately. In order for this to be a question of a meaningful activity, i.e., an activity governed by a goal – and only such activities are permitted, according to Dingler, in the construction of exact sciences –, the goal of the action must be defined; thus an exact definition of 'periodic' is needed. Meanwhile the 'strict identity of all essential conditions' is not such a definition, as long as it is not stated what conditions are 'essential'. Dingler wisely avoided requiring the indistinguishability of individual periods in all conditions at all formulable, something which would obviously be nonsense. However, the problem of distinguishing certain conditions from 'inessential' conditions justly remains.

The procedure which Dingler may have been thinking of in this context ("in that namely new circumstances are discovered which can influence the course of a process . . .") is nevertheless circular. In order for such a discovery of influences to be at all possible, there must be at least crude realizations of periodic motions, i.e., what 'periodic' means must already by defined.[14] Dingler's definition of periodic is an empty requirement and consequently there can be no realization. Furthermore, this definition cannot serve to interpret the practice of clock building.

Even if one sympathetically views Dingler's conception of the foundations of chronometry as one merely needing completion, nevertheless the question of how the presence or absence of distinctions of individual periods (even in different places) can at least be noted if it has not already been verified in established procedures remains unresolved.

Here, so one might suppose, Dingler has fallen victim to the remarkable precision of clock technology and has sought a theoretical underpinning of the watchmaker art that is all too close to practice. Objections apply to his arguments which objections are analogous to those applying to the official doctrine of physics on the topic of time-measurement. Dingler adjoins a noteworthy discussion about the basic concepts of dynamics to the definitions which produce basic concepts of geometry and chronometry or at least should produce them. Since these, in accordance with the methodological remarks in Section 4 of this chapter, are likewise incomplete – here, however, we present to proposals for solving the deferred problems –, they shall not be reviewed here. A thorough elaboration of the definition of mass by the law of gravity, contained in Dingler's later works, and an application to the foundational questions of astronomy is presented by Thüring, 1967.

3. LOGIC AND PROTOPHYSICS. ON THE FOUNDATIONAL THEORY OF PAUL LORENZEN

Dingler's philosophy of science has been taken up and developed further by Paul Lorenzen. The works of Lorenzen dedicated to this theme are more easily surveyable than those of Dingler and, in particular, are written in a style which does not require any historical or philological feats of interpretation. Therefore a compressed presentation is in place, one which, above all, emphasizes the points that go beyond Dingler, but one which also makes clear some differences with the view expressed in this book.

For Lorenzen the problem of valid or true universal statements does not become relevant first with the question 'How is physics as a science possible?'; rather it already arises in general form at the most elementary level of 'rational' speech, i.e., speech that constantly strives to make good its claims, and thus belongs to logic in the wider sense. The superiority of Lorenzen's philosophy in this respect vis-à-vis that of Dingler's may be briefly characterized by the fact that it overcomes the linguistic-philosophical and logical naivêté of Dingler.

The possibility of defendable speech (in the sense of 'defened in a dialogue if questioned'), in spite of all the ambiguities of ordinary language, can be secured by means of a reconstruction of language that extends from the most elementary building blocks up to the high stylization of scientific terminology. (The assertion made in the analytical theory of scientific language that theoretical terms are only partially interpretable does not contradict this assertion. Here it is not a question of the interpretation of presently existing physical theories, but of the terminological construction of theories that are alternatives to these.) It may be taken as a criterion of scientific method (*Wissenschaftlichkeit*), that the requirement of explicitly demonstrating the validity of propositions, from the use of words up to the criterion of validity itself, can always be fulfilled. The question which immediately raises itself here then is the question of how the above-mentioned reconstruction is to proceed.

Among publications concerned with an answer to this question, the

Logische Propädeutik[15] of W. Kamlah and P. Lorenzen is the most compelling types of steps which are to be taken in the individual stages of the reconstruction, we shall investigate the legitimacy of such an undertaking since it indeed claims to be a methodological beginning to exact speech.

The propositions representing the heart of the 'Logische Propädeutik' are not propositions asserted with a claim to truth – in whatever sense –; rather they must be understood prescriptively. The individual steps to be recorded in the following are given as rules. Consequently a catalogue of rules for the reconstruction of rational speech is a methodological beginning which requires no foundation of its truth. At the most, rules can be followed or ignored; a decision for or against them must be argued, when occasion arises, morally in each individual case.

From the level of an exact language already set in motion, one may argue compellingly for the fact that every methodologically ordered construction of a language starts with elementary predication. J. Mittelstrass states this fact as a resumé of an investigation of 'Predication and the return of the identical' (Mittelstrass, 1968) as follows: "The trivial and almost paradoxically sounding statement that we cannot speak about the world without (linguistic) distinctions, turns out once again to be true. Expressed in a more sophisticated way this means that all reflection about predication makes use of that very predication and is thus incapable of yielding any kind of foundation upon which predication might be based. As a basic linguistic performance predication cannot be circumvented". (*Loc. cit.* p. 87.)

The 'predicates' which are asserted or denied of objects designated by proper names can be introduced 'by way of example', i.e., learned from examples and counterexamples. A study of finitely many (*de facto*, mostly very few) examples and counterexamples implies, to be sure, a certainty which at most satisfies the needs of daily life. Borderline cases are not considered and it cannot be precluded that a case always will appear at some time, in which case differences of opinion with respect to a predicate are only resolvable through a new agreement concerning even the case in dispute.

'Predicate rules' attain a first exact definition of speech; such rules, as the functioning of unreflected ordinary language may be understood, hold implicitly there; however they must be explicitly agreed upon in a scientific language. Once they are established, they form a foundation of material, i.e., not formal, inferences. Thus, for example, the inference from 'x is a man' (where x is to be replaced by a proper name) to 'x is mortal' is to be considered as valid for the reason that in our speech community all men are designated as mortal, something which has the rank of a purely linguistic convention. To be sure the justification of such predicate rules is no longer a matter of logic in the broader sense. A closer clarification of the relationship of predicate rules to their basis of legitimitation – in the analytic theory of science this is designated as the adequacy of predicate rules – is found for the first time in the book written together with O. Schwemmer, *Konstruktive Logik, Ethik und Wissenschaftstheorie*. Predicate rules, indeed, do regulate explicitly the use of predicates and have in precisely this sense the character of conventions; but with this it should not be asserted that they are 'mere' conventions, i.e., that they have no assignable reasons for their conventional establishment. Predicate rules, rather, correspond to a prohibition norm based upon linguistic acquisition and use in the life-world. This is illustrated in a rule proposed in Chapter III:

The rule 'If a body K_1 is moved relative to a body K_2, then K_2 shall also be said to be moved relative to K_1' is not intended as a categorical imperative to the effect that whoever asserts the motion of a body A relative to another body B, ought now also make the corresponding symmetrical statement. Rather the rule signifies that attacking or doubting the statement 'K_2 is moved relative to K_1' is prohibited if the statement 'K_1 is moved relative to K_2' has previously been accepted. This prohibition norm does not receive its validity from a mere convention in the sense that another prohibition norm, e.g., one implying a one-place or non-symmetrical use of 'moved', could have been chosen equally as well; rather it receives it from the meaningful (i.e., suitable) and proven practice in the life-world of using 'moved' as a two-place predicate, which becomes a symmetrical one by means of the maxim of not distinguishing reference bodies in a science of bodies and their motions, where the maxim is likewise based within the life-world. This example shows that already the explicit standardization of the use of predicates by rules cannot succeed without recourse to the life-world in which practices (of a linguistic as well as of a non-linguistic sort), experiences and maxims of conduct decide the recognition of a prohibition norm implying predicate rules.

Thus predicate rules attain an exact definition of speech through instructions as to how one is to go from one proposition to another with words that have already been introduced. New terms not subject to the vagueness of definitions by example are introduced by 'definition' proper. The form of such definitions is $x \in P \leftrightharpoons A(x)$, i.e., the term to be defined, 'P', is asserted of an object 'x' if the proposition $A(x)$ holds for 'x'.

Predicate rules and definitions are easily distinguished from one another: The former establish conventions which exactly define the use of at least two predicates among which, as a special case according to the just cited example, are to be reckoned many-placed predicates with different occupations of the variable places. Their inference direction (go from $x \in P$ to $x \in Q$!) is univocal. Definition, on the other hand, fixes only one term and logically considered is an equivalence relation. (A more basic treatment of problems in the theory of definition is found in Gabriel, 1972).

The construction of a scientific language, in which, after predicates have first been fixed by means of rules, terms are then introduced per definitionem on this basis, cannot manage without a further class of terms; these terms are called 'abstractors' in the *Logische Propädeutik*. (Logical) Abstraction, which consists in ignoring irrelevant distinctions of individual objects, leads to abstractors, which are words for 'concepts'. Here 'concept' is to be understood as an abstraction from word-sounds (*Lautgestalt*). Along with the predicate 'red', it is meaningful to speak of a concept 'red' which is invariant with respect to synonymous designations – say in another language. Finally, for the systematic reconstruction of the technical language of physics, in particular, yet one further group of terms is necessary; this group (in analogy with the technical term 'abstractor') shall receive the designation 'ideator'. We will speak in yet more detail about the role of such ideators (for example, 'plane', 'point', 'uniform motion'), when we now discuss P. Lorenzen's view of the methodological foundation of physics.

If we start with the fact that the endeavors in the *Logische Propädeutik* have reached a degree where all differences of opinion regarding the use of individual words are resolvable in principle, then the next question focusses on justifying propositions. We shall take this up here, merely in overview, by elucidating the pyramid of concepts found repeatedly in Lorenzen.

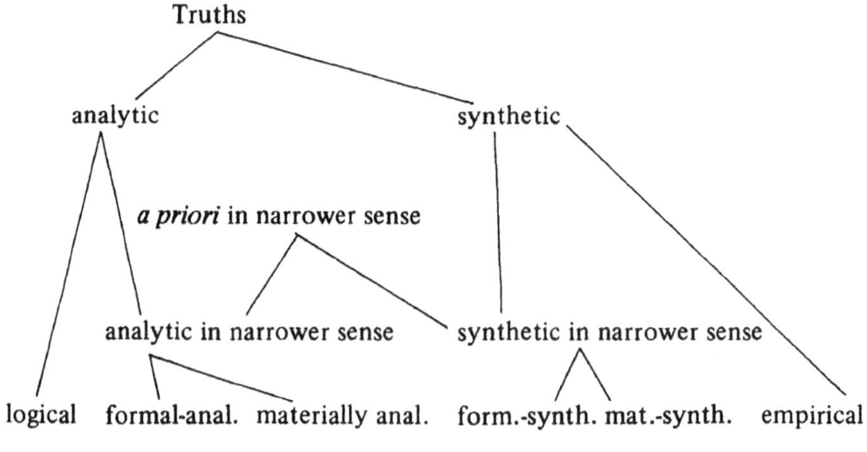

Of the six terms standing at the base of the pyramid, the fifth one – material-synthetic truths – in particular will be of interest here; the others will be mentioned only in passing.

Analytic truths fall into logical truths, which hold tautologically, i.e., by reason of their form alone, and into analytic truths in the narrower sense. Under these fall formal-analytic propositions, which are true by reason of logic and at least one definition, as well as material-analytic propositions, whose validity also depends on at least one predicate rule.

Both may be called *a priori* in the narrower sense, in the same way as the two groups of synthetic truths in the narrower sense. Formal-synthetic propositions (for example, arithmetical propositions) require, for their justification in dialogue, a recourse to the construction rules of the objects discussed. Material-synthetic propositions contain ideators and can be justified only through reference to 'ideative rules'. The domain of material-synthetic propositions is determined through a theory of the foundations of physics. We will now discuss this.

According to Lorenzen, physics is 'universal somatology', i.e., a theory of bodies which is free of proper names for individuals, i.e., a theory of bodies "insofar as they are nothing other than bodies". (Lorenzen, 1974, p. 143.) Thus individual somatologies, such as astronomy or geophysics, do not belong to it.

Here the locution 'bodies, insofar as they are nothing other than bodies', deserves special consideration. The introductory example, which depends on Aristotelian ontology, namely the descending sequence man-animal-plant-stone was given in order to obtain, with stones, specimens of objects which are nothing other than mere bodies, but it immediately provokes the question of what properties in the other groups are to be disregarded insofar as they are considered only as bodies. The answer that, in contrast to stones, these three are animate leads to a task which in my opinion is undecidable by physics in particular, namely the task of defining 'living' or 'life'. It was to escape this dilemma, however, that in the work cited recourse was had to the simple Aristotelian example. (This difficulty is first resolved in P. Lorenzen/O. Schwemmer, *Konstruktive Logik, Ethik and Wissenschaftstheorie*, p. 123, according to which changes of bodies not completely describable with the descriptive means of physics are named as a criterion for living. A far less problematic definition of physics than may be given by relying on categories of Aristotelian ontology is found likewise in Lorenzen-Schwemmer, where, per the heading, physics is treated under the 'theory of technical knowledge'. In this context, why physics starts with geometry would, then, indeed have to be explained in addition.)

A consistent interpretation presents itself, then, if the definitions of 'body' and 'physics' do not claim for themselves final rigor, but rather if they serve more for a preparatory discussion leading to protophysics – which is then the rigorous theory. Introduction by way of example of 'living' suffices for this. Presystematic knowledge that omits peculiarities, as for example in the case of an animal when one disregards the fact that it is living, may achieve a terminological determination of 'body' and consequently of 'physics' sufficient for a preliminary orientation. Should 'body' be available as a term of protophysics itself, it would need to be made further terminologically precise. (Cf. p. 101)

The so-called 'laws of nature' – what status they occupy in the strictly methodological construction of physics has yet to be presented – are quantitative propositions; they make use of arithmetical language to represent measurement results. The question of how the objectivity of physics is possible, consequently leads to the further question of how objective measurements are possible. An answer to this question which, "considered from the viewpoint of mathematics, contains the first steps that must be taken before one can employ physical measurements", (Lorenzen, 1974, p. 148) is given by 'protophysics.'

The systematically earliest measurement is that of length; all further measurements, for example of time, mass, etc., already require measurement of length. (In Lorenzen/Schwemmer as the reason for beginning protophysics with geometry is given the "factual genesis of measurement" (Cf. p. 162); i.e., the reason given is not a method(olog)ical one as is explicitly the case, and with reasons given for it, here in this book (Cf. p. 69). So then Lorenzen's main interest focusses on geometry, which ought not to be considered as a merely formal theory – its propositions should also hold for real bodies –; rather, in its intermediate position between formalsynthetic and empirical truths, it requires its own distinctive foundation. Since geometries already present themselves in logically strict, axiomatic elaborations, this thus becomes a question of a foundation of axioms.

Proving their validity cannot be undertaken before the words 'plane', 'straight line', 'orthogonal', etc. which occur in them are exactly defined. Lorenzen took up Dingler's proposal to define geometric basic forms by means of indistinguishability requirements – or more strictly since here it is not a matter of a definition in the given sense (See p. 53) – to determine them 'ideatively'.

What is of significance for 'ideative definition' is presented for the first time in detail in the *Logische Propädeutik*. Both earlier works (Lorenzen,

1961 and 1964) still state that in the formalized representation of the so-called 'homogeneity principles' of the plane – a representation strictly formulated in accordance with the Leibnizian conception of the indistinguishability of two objects – 'arbitrary' 'geometric' propositions can occur. Thus, if the plane should be distinguished from other surfaces by nothing that (along with the indistinguishability of half-spaces divided by the plane) its points are indistinguishable – which is formulated as the substitution rule $A(E,P) \to A(E,P')$ – then the proposition A can "be arbitrarily constructed with the help of the logical particles" from the basic propositions of the theory. There are:

$$P \epsilon E, \quad P \epsilon g, \quad P_1 E P_2, \quad E_1 \| E_2, \quad g -| E$$

(Point P lies in the plane E; P lies on the line g; E lies between P_1 and P_2; E_1 is parallel to E_2; g is perpendicular to E.)

So long as a rule

$$P \epsilon E, P' \epsilon E, A(P,E) \Rightarrow A(P',E)$$

(Lorenzen, 1974, p. 133) is formulated in the metalanguage, this replacement of proposition A is possible. A does not contain any free variables. The object variables P, g, and E appearing in the basic propositions of the theory are treated as if they designated well-defined objects. The transition to the axiom schema

$$P \epsilon E \wedge P' \epsilon E \wedge A(E,P) \to A(E,P')$$

($\wedge$ denotes logical conjunction, $\to$ implication)

meanwhile brings no profit as long as no possibility exists for forming a proposition from this propositional form. And the basic propositions of the theory are not available for this, since the objects occurring in them (planes, lines, and points) are only determined via these axioms, thus not already by means of the axiom schemata.

This difficulty is partially resolved in the *Logische Propädeutik*. After 'sides', 'edges' and 'corners' as well as 'planar' (for sides) and 'straight' (for edges) are defined by example as distinguished forms, propositions – let it be noted, propositions always still burdened with all the vagueness of merely examplary introduction – about orthogonality, parallelism and complex composed geometrical, more accurately 'pre-geometric', relationships may be formed. However, it is only these pre-geometrical propositions – which thus treat of vaguely determined objects, such as, for example, exemplarily planar sides – which can be substituted into the axiom schema. A plane defined only by example becomes, then, a realization of an ideatively defined plane, when all its vaguely defined points are indistinguishable in respect to all pre-geometrical propositions. (Strictly considered indeed even the pre-geometric vocabulary given in the *Logische Propädeutik* does not suffice. It must be extended to words that are needed to draft instructions for producing smooth surfaces on bodies. A complete listing of this vocabulary is offered for example in Janich, 1976a.)

Here it becomes obvious how a definition of this sort deserves to be called 'ideative'. The realization of a basic form can be certainly always better accomplished, though never completely. While Lorenzen, in 1961, wrote that the question of realization serves "only to make clear homogenous basic forms" and that "later logical analysis and derivations from the homogeneity principles will be independent of questions of realization" (P. Lorenzen, 1974, p. 130) – a view which agrees with Dingler's – this account becomes only fully intelligible when the statement in the *Logische Propädeutik* is considered: "We will see that geometric norms are nothing that would be 'separate' from our procedure for manufacturing geometric devices – it is only that they are not 'realized' in our devices (and, in contrast to Aristotle, also not 'abstracted' from these devices) – they must rather, as Plato said, be known *before* the manufacture of the devices". And: "It is to be kept in mind that ideative norms are instructions for action". (Kamlah and Lorenzen, 1967, p. 229.)

Insofar as ideative norms formulate no expectations of objects of nature, rather contain instructions for producing instruments, it may be asserted that laws are *prescribed* for nature. Natural objects are worked upon, which means, within the scope of geometry, that ideatively determined forms are forced upon bodies by working upon them, which simultaneously will guarantee the validity of theorems derivable from ideative requirements.

Finally it is the goal of protophysical geometry "in particular to define *congruence* with the basic concepts that have been defined by means of ideative norms", from which "in any case the irrefutability of geometric theorems by means of length measurements" (Kamlah and Lorenzen, 1967, p. 231) follows. Notwithstanding, however, the fact that up to now no complete foundation of a geometric axiom system by means of homogeneity principles exists, important questions are still open:

1. Lorenzen defined the plane by means of two homogeneity principles, i.e., by means of an internal one that requires the indistinguishability of the points of a surface and consequently is also satisfied by spherical surfaces,

and by means of an external one that requires the indistinguishability of the half-spaces divided by the surface by requiring the indistinguishability of all points not lying on the surface. This external homogeneity pricincple raises the question of what the propositions are relative to which the indistinuishability of the points not lying in the surface must be realized, of how these propositions are connected with the realization procedure for plane surfaces, and, above all, of why such an external homogeneity principle would at all be formulated, since plane surfaces may also be defined (and thus more economically) as surfaces that are internally homogeneous but neither concave nor convex. Besides the connection between an external homogeneity principle which postulates the undistinguishability of all points laying at both sides of the plane, and the realization proceduve cannot be seen. To the external homogeneity principle it may be objected materially (even if this in fact occurs mostly for methodologically circular reasons) that it implies the Euclidicity of the geometries that build upon it, without the external homogeneity principle being argued for – on this point Lorenzen only suggests that his account reconstructs the actual genesis of geometry and mechanics.

2. Lorenzen, to be sure, argues within the framework of an ethically grounded reflection that goes beyond the foundation of geometry for reconstructions of the type proposed by him; but he does not state in detail the ends that are pursued with particular homogeneity principles, such as the external one of the plane. Thus – without lapsing into Dingler's old error – the reason for the Euclidicity of the geometry striven for must be stated (which reason in fact may be stated, namely e.g., the uniqueness of the length, measurement that is striven for, in the two-fold sense that the basic forms planar and orthogonal are uniquely reproducible and that congruence is demonstrably an equivalence).

3. Still to be clarified is what role is intended for the homogeneity principles in Lorenzen, since in all homogeneity principles previously formulated for orthogonality (in the *Logische Propädeutik*, for example, for two planes, in his latest works for a straight line and a plane) there already occurs an assertion of orthogonality. Thus it cannot be thought, in analogy with the internal homogeneity principle of the plane, that a given plane (a given straight line) for whose points indistinguishability requirements have been formulated precisely thereby has been determined ideatively to be orthogonal. Rather there already occurs a predicate 'orthogonal' that may be established, by example or by definition, in procedures for producing a 'straight edge' (cf. the preface to this volume). The appearance of this predicate in homogeneity

principles for orthogonality thus gives rise to the appearance of circularity, which would still have to be dispelled.

4. Finally an (admittedly gradual) distinction between Lorenzen's works on protophysics and my own works must be mentioned, a distinction which possibly will get lost in the further elaboration of the statements under consideration, but one which at present entails differences of opinion on particular points: The reconstruction efforts of Lorenzen are oriented more towards the reconstructing formulas to be obtained than towards the pre-scientific realization possibilities. The realization procedures and their role in the formulation of very definite homogeneity principles is scarcely discussed in Lorenzen – as was already mentioned above in the case of the external homogeneity principle. In this he stands closer to Dingler, for whom, likewise, it was more a question of an 'intellectual' reconstruction than of an operational foundation in the strict sense: It plays no central role whether the given operations 'really' can be carried out concretely (and perhaps supported by the history of technology). One of Lorenzen's later works, (Lorenzen 1975) in which the definition of mass is developed, clearly expresses this: Lorenzen proposes to define the ratio of two masses by the reciprocal of their velocity ratio before an ideally inelastic collision (referred to the combined mass after the collision). Thereafter the law of momentum holds *per definitionem*, as does the law of energy for the elastic collision. For the question of realization this merely means that realizing ideal inelastic collision events (*Stossvorgänge*) implies so carrying out the (central) collision of two bodies that relative velocities before and after the collision would be independent of the magnitude of the velocities. It is not stated by what actions this independence can be attained.

Suppose that on a craft level it has been stipulated with a pre-scientific vocabulary how a central inelastic collision of two bodies is to be carried out (let all velocities be measured relative to the earth); the ideative condition of Lorenzen will now be formulated: the velocity ratio of the bodies before the collision, which ratio is referred to the combined mass, should be velocity invariant. Here remains open what a physicist should do if he measures, say, deviations in the velocity ratio increasing in one direction with increasing velocities. One could answer, in the spirit of Dingler, that the physicist must then search for disturbing circumstances, that from this he must seek exhaustion knowledge about relevant accompanying circumstances. Against this, however, (and, in contrast to objections based on relativistic considerations, non-circularly) it may be said that the *expectation* by velocity invariance is founded by nothing, even if the end of obtaining a velocity invariant definition of mass ratio is justified. In other words a realization procedure is lacking, the description of which logically implies velocity invariance. Only when, by means of instructions for action (Handlungsanweisungen), the technically controllable components of a state of affairs (here: of a collision event) are explicitly clear, can deviations from it be traced back to disturbing circumstances. Lorenzen's definition of mass ratio is thus not operative, indeed it ignores precisely the central aspect of protophysics, i.e., that physics builds upon the technically and theoretically controllable properties of artefacts (to which, here, in addition to apparatuses, are also reckoned events in instruments).

This objection, based here upon a definite conception of protophysics (cf. the role of uniqueness proofs for the definition of disturbances p. 00), also may be formulated differently, and what is more without recourse to physical theories as well as without reference to general principles of protophysics. Without a doubt Lorenzen would require uniqueness of his definition in precisely the sense that two given bodies K_1 and K_2 have *exactly one* mass ratio, i.e., that there is one and only one rational number which gives the velocity ratio of both bodies before the collision (*ceteris paribus*). Moreover, in the mass definition no proper names (e.g., 'earth') occur. If the event assumed by Lorenzen were to be at all suitable, three velocity measurements relative to a body K_3 would have to be performed, namely of K_1 and K_2 before the collision and of K_{1+2} after the collision. Otherwise reference to the combined body K_{1+2} is not possible. Artisans and, as a rule, physicists choose the earth as K_3. Since, however, this choice should play no role, a body 'strongly accelerated' against the earth must also come into question as the reference body K_3. Referred to this body, however, K_1 and K_2 have a different mass ratio than with reference to the earth, whereby 'strongly accelerated' is relativized to the procedure of measuring velocity actually employed at the time. In other words, for each determination of a mass ratio actually performed, another reference body can be given which yields another mass ratio for the same bodies. Thus for the definition of mass by means of an inelastic collision, there arise problems very similar to those known from classical mechanics, more specifically, problems of definition for inertial systems.

The objections presented here affect two not yet completed theories, geometry and hylometry. ['Hylometry', a term proposed by Lorenzen, is formed analogously to the terms 'geometry' and 'chronometry', and means a protophysical theory of measurement of mass.] Untouched by this is the fact that the works of Lorenzen on a constructive theory of science in the areas of the theory of scientific language, logic, mathematics and protophysics represent, at present, the sole visible chance of escaping an affirmative theory of science that lags unlimitedly far behind the factual development of physics; they also represent the sole visible chance of coming to grips with the natural sciences with a philosophically critical self-understanding. If therefore Lorenzen himself says, "the critical business of a constructive theory of science for physics will be postponed temporarily" (Lorenzen and Schwemmer, 1973, p. 255) – a dictum in which it is not clear whether it concerns an assertion or a command, or whether it holds only for Lorenzen himself or also for others – one should categorize this sentence as merely rhetorical and take it less in earnest.

4. ON THE METHOD OF PHYSICS

In the following, one of the proposals of Lorenzen's continuing methodology shall be developed as far as is required to elaborate thereby a theory of time measurement (Chapter III) and at the same time to develop the alternatives to the views described in Chapter I. For this purpose, as was already announced in Chapter I, we will set out from the physics of the physicists. This physics, however, shall not be analyzed solely with the help of its theories or history of theories and in precisely this sense be understood only 'technically" as in the affirmative theory of science, rather it will be analyzed in its claims. Here, on one hand, we will explain more precisely what it means for physics to be an empirical science of nature and, on the other, how claims to scientific nature, objectivity, validity (such methodological terms are to be defined exactly) may be understood. A technical understanding of physics is presupposed for these questions. In addition to this, a procedure for securing the validity of individual measurement results will be given for the special area of general metric terms.

4.1. *Physics, Natural or Experimental Science*?

What physics is, is determined prima facie by physicists. In the scientific workday, to be sure, the task 'of determining what physics is' arises increasingly seldomer, since in physics textbooks such definitions are increasingly abandoned. If need be in preambles to regulations for the conduct of an examination or in demarcations of scientific organizations from other professions, physicists themselves still speak today about physics; otherwise it is practiced. Thus today there is already something anachronistic inherent in efforts at defining and delimiting an individual discipline such as physics. They can, if necessary, still be considered in an encyclopedic aspect; but even there they are quickly suspect of furnishing a scientifically systematic knowledge of doubtful use and little proximity to scientific praxis. Here we too shall refrain from a 'definition' of physics, which would appear with some sort of claim to completeness. It is taken as self-evident that the term physics signifies not only a corpus of theories, but also a scientific concern to which belong the art of experimentation as well as the institutions in which physics is organized in and outside of universities, including the rules of the game of physicists, such as publication style, foundations for honoring achievements in physical research, a normal self-understanding on the part of physicists and certain regulations for the education of the coming generation. In order to suggest thereby that 'physics' (as in ordinary language) is employed as a quite vaguely defined collective term for an abundance of different components of aspects of an historical phenomenon, we shall speak of 'historical physics'. Historical physics is thus the result of a special history, which one today generally knows is not only a history of theories.

Making historical physics itself a subject of scientific research is a very costly venture to be divided among many special branches of sciences; it is a venture that, indeed, today has been started from many different points, but which still lacks a more exact definition of its goal. Where such scientific research is not borne by the false empiricistic pathos that one must first do as much empirical investigation as possible (without quite knowing to what end) in order, then to have knowledge whose utilizability would then suggest itself, i.e., where scientific research of precise questions is pursued to the solution of standing problems, there is to be expected of such research, surely, a picture of science that is better than the traditional one. If, therefore, we choose another path here, we do so neither as an undertaking in competition with a goal-oriented scientific research nor is our choice accompanied with a verdict against an empirical-scientific coming to grips with physics. We must merely reject here a claim of exclusiveness, according to which no investigation of physics other than an empricial-scientific one could arrive at results relevant for physicists. (Cf. Janich, 1981.)

Notwithstanding a wealth of further possibilities of defining historical physics, we can state two minimal requirements here without great empirical or historical research; these are requirements that physicists do not reject – neither in fact nor by implication – namely the requirement that *physics be a natural science*, i.e., in a sense to be more closely determined it is concerned with *nature* and that, in a sense likewise to be defined more exactly, it is *scientific*. These minimal requirements, which again are problematic in almost every formulation today, may be agreed to at least *ex negativo* as generally recognized: professional physicists would agree neither that they practice a formal or ideal science such as mathematics, or a cultural science such as history, nor would they acknowledge that the arbitrary is allowed in physics, i.e., that no rules of any sort are valid. (For the dependence of the subject matter of physics on those methodological claims see: Janich, 1977b.)

In the following, now, both these claims of physics to be a science and to be about nature will be defined exactly to the extent that, on one hand, it can still be expected that each physicist would assent to them and that, on the other, they will allow one to ask by what means these claims will be made good. Proposing such means in the form of a methodology, as well as redeeming this methodology with the example of a theory of time measurement are then the partial efforts of a so-defined '*rational reconstruction*' (or synonymously) 'foundation' of physics. Let it be expressly pointed out with what weak premisses, namely in the form of factual statements about historical physics, the task of reconstruction can be formulated: here only two (besides possibly many others) necessary criteria of physics (necessary in the sense of indispensable minimal requirements) are resorted to; again, in light of them, necessary methodological presuppositions are merely stated, and this still with the reservation that there could be other equally good proposals for this. Thus there can be no talk of protophysics (as a partial result of a rational reconstruction) being bound to a given 'paradigm', e.g., that it would be oriented toward classical physics.

Now to define both criteria exactly. If physical theories already could be drawn upon for this, we could say much, and it would be terminologically exact, about the object domain of physics. Trivially, however, such definitions are after the fact; they depend on the historical state of physics and the current recognitition of theories and consequently are not permissible in an endeavor to inquire critically whether the present theories or the methodology underlying them can make good the claim to be knowledge of nature. On the other hand, no one can participate in the debate being conducted here who has not already learned to speak and act, who has not already learned on a 'life-world' level what it means actively to cause something, in particular to produce devices oneself or to employ those produced by others. It can thus be expected of anyone who here possibly would like to formulate an objection, that he can distinguish between what he himself has actively changed and what he has found given. Even though it still really can become a problem of a highly developed empirical science, whether the given (*das Vorgefundene*) perhaps has not been changed already by other human agents, nevertheless everyone can understand what it means to speak of the given that has not yet been actively changed – not by the person who encounters it and not by other men.

The given, which, above all, could be exemplified in astronomical objects, but also in earthly bodies and events, need not be determined more closely for our purposes. We can dispense here with distinctions such as living-dead or similar ones that occur in traditional definitions of physics. (Further specializations arise namely 'of themselves', i.e., through selection of particular experimental techniques and linguistic means.) What remains crucial is only that basic human abilities, namely being able to speak and act, are presupposed here – and what theoretician of science or philosopher, let alone scientist, would not presuppose that?

With the help of the human ability to act and to talk about these actions (e.g., their purposes, their results, etc.), the distinction *'natural-artificial'* can be introduced. *'Natural'* means what has not been changed by man; *'artificial'*, what is the result of human activity. Thus apparatuses, but also events caused by men, are called artificial. Associated with this distinction are two traditional goals of physics that are repeatedly mentioned but which in their connection, until today, have not been fully transparent: The prediction of natural events and the technical production of effects (as an indirect activity, namely one using instruments) with apparatuses.

Regularities can be observed in the naturally given, they can be induced in instruments. It would be surely absurd to postulate a distanced, disinterested observer of nature (in the sense of natural objects and events). Observations (even now: without employing instruments such as telescopes, clocks, etc.) are made with a purpose in mind; a farmer and a sailor observe different regularities in the weather. The (instrumentless) art of prediction as an inductive extrapolation of observed regularities depends on empirical knowledge that has been collected in a goal-directed way. Since physics has long left the state of instrumentless observations, in the case of the observation of natural objects and events with instruments both natural (i.e., only in their selection, oriented toward human needs) regularities and technically enforced regularities occur. The greatest part of present-day experimental physics, however, observes only artificially changed objects or artificially caused events. To that extent the designation 'natural' science is just as misleading as 'physics' (stemming from φύσις and reminding us of 'the born' or 'the grown').

It thus appears more suitable today to speak of an *experimental science*, instead of a science of nature, i.e., to characterize physics not by means of the domain of objects investigated, but rather by means of a domain of methods. Presumably one also would have the assent of any physicist herein that physics is an experimental science, which does not exclude the fact that the natural is also observed. (For the relation between observation, measurement and experiment, cf. Janich, 1978b.) This shift to characterizing physics as an experimental science, instead of as a natural science, because the latter characterization is misleading, thus plays a decisive role; it does so since according to it physics consequently no longer primarily has to do with natural properties and events but rather with artificial properties of instruments and events, i.e., those produced by man, and also the artificially produced and sustained regularities.

It is only repeating what is known to say that physical experiments, for the most part, include measurements, and that even where an experimenter does not seek a result in the form of a measurement value, initial conditions of the experiment must be quantitatively established. In the indispensable constituent of modern physics of being able to make quantitative, reproducible experiments, the domain of the natural has been wholly abandoned: every quantity is artificial. Even the observation of quantities in natural objects or occurences presupposes the manipulation of other objects and can only come to light through reference to the results of manipulation (e.g., the diameter of the moon, which certainly has not been changed by man, first comes to light in an observational procedure using artificially produced instruments, in which instruments it is precisely the artificially produced properties that are decisive for the observation result. *A fortiori* this holds for the observation of artificial objects or occurrences in 'experiments').

The claim that physics distinguishes itself, e.g., from mathematics or history because it is concerned 'with nature' can thus only be understood without the aid of physical theories in the sense that physics is a quantitative experimental science. With this we have entered upon the path to a methodological, but let it be noted, still 'paradigm-invariant' characterization of physics; this path will now be pursued further in order to be able to clarify the second claim, namely of being a science.

4.2. *The Claim to Scientific Nature* (*Wissenschaftlichkeit*)

In the sense of minimal requirements, it is again trivial to assert that physicists in fact raise a claim to be following scientific method, although it must be questioned whether this scientific nature would be characterized unanimously according to the same criteria. Quite apart from further and sharper definitions, physicists – and this again with good reasons and generally admitted – would certainly assert of their research results in the form of theories that these results are distinguished from mere opinions, because they have been formulated in a thoroughly standardized language and tested by experience, where by experience is intended the experiment that is in principle repeatable.

One thus finds oneself in agreement with professional physicists, and with almost all philosophies of science in awarding a special status to the statements of a technical science. Subject to further exact definitions, the claim to scientific nature shall consequently be understood as a claim to *'objectivity'* or *'intersubjectivity' of physical propositions*. In the word objectivity there is expressed an older misunderstanding that is encountered scarcely anymore since the linguistic-philosophical investigations of Wittgenstein, Russell and the philosophers of the Vienna Circle, i.e., that the objectivity of natural science is due to its 'objects', namely the natural objects and processes governed by laws. (Only in the naive belief that experiments are repeatable because nature is ruled by laws does one still encounter the remnant of this concept of objectivity.) Thus here we shall be concerned only with the intersubjectivity of physical propositions as at least a necessary criterion for the scientific nature of physics. As the history of controversies in the theory of science teaches, it is – cautiously formulated – dangerous to speak of physical propositions without further differentiation. In light of the aim of the present investigation, therefore, all physical propositions in which at least one universal quantifier appears will be excluded, i.e., all physical laws (and with this, we shall not suppress it here, above all the propositions that are 'interesting' for physicists as well as for theoreticians of science.) There remain, thus, above all propositions about particular measurements results, i.e., propositions such as: This table-top has a length-width ratio of 0.65. The diameter of earth and moon stand in the ratio 12,742:3,476. (Here, then, the use of units such as the meter is also dispensed with.) What does it mean for such a propositions to be intersubjectively valid?

As is well-known, the question of the intersubjectivity of such measurement protocol sentences is not treated in isolation in other philosophies of science, rather the view has prevailed that even the simplest propositions occurring in physics or the operations underlying them are theory dependent. As shown in Chapter I, this view is due to the fact that one started with the analysis of complete theories of physics existing at present. The contrary view advocated here can be examined in the following. There it will be demonstrated in detail how one, step by step, arrives at intersubjectively verifiable measurement protocol sentences. This also can furnish a reply to the opinion that, for physics, one must assume a non-statement view, since physical propositions are always to be relativized to certain groups of researchers as well as to intentions of applications. If here, on the contrary, the older claims will be maintained, i.e., to pursue further the intersubjectivity of physical propositions and to seek for possibilities of securing them, then it cannot be objected that this is not possible for man (Cf. Stegmüller, 1973, p. 26, where Stegmüller speaks against the human possibility of foundations postulates similar to those advanced here) – in any case, any proof of this has been lacking up to today; even the beginnig of a proof is lacking. That analysis of actually existing theories in a particular direction in the theory of science has led to the result that the claim to intersubjectivity in the form formulated more strictly above – i.e., without the mentioned relativization – has not been made good, can obviously not be taken as such a proof. Here, namely, for one, the maxims guiding such an analysis are questionable; secondly, from an actual development there does not follow the impossibility of another one; and thirdly for the factual development of physics it must be clarified first whether or not, in the laboratory and theoretical praxis of physicists, claims are fulfilled which do not recur in the non-statement views.

The intersubjective or, synonymously, interpersonal verifiability of measurement protocol sentences is dependent for one thing upon the standardization of the linguistic means employed. Here we must clarify, going beyond the procedure applied in physics in an extremely unreflective way, how such standardization must take place. Physicists, however, not only point correctly to a far-reaching normalization of their language – above all through the employment of mathematics – but they also distinguish their experimental science from, e.g., historical science, which historical science likewise rests on given experiences, by means of the claim of the repeatability of experiments. Restricted to measurement protocol sentences, this implies the repeatability of measurement. By this it is obviously not meant that the measured (natural or artificial) object has remained the same – one does, indeed, take measurements to decide, among other things, this question –; rather by it is meant that the measuring instruments have remained the same, in crucial properties, i.e., that relevant circumstances are the same whether 'of themselves', which one then only need verify, or because of efforts. For the special case of measurement protocol statements this means that ***the properties of measuring devices must be reproducible***.

To identify the repeatability of measurements and the reproduceability of measuring instruments lead to a critique raised by some commentators on protophysics (Cf. Böhme, 1976b; Fertig, 1978; Hucklenbroich, 1978). The objections stress the fact that protophysics cannot determine the structure of the objects measured by means of norms guiding the construction of measuring tools. In fact protophysics never claimed to do that. To the contrary, the decisive progress made by protophysics lies in the shift from arguments drawn from the objects of scientific experience in an unclear and often ontological way towards the methods to gain reliable knowledge about these objects. Fertig's additional critique, drawn from his model theoretical approach of measurement, that protophysics does not distinquish between the construction and the application of measuring tools is explicitely wrong (as may be seen already in earlier writings on this topic; cf. Janich, 1973).

In laboratory practice the control of properties of measuring devices, and that is justified by the ends of this *praxis*, ensues with the help of other measuring devices and by consulting knowledge of physical laws. These procedures of employing measuring devices however, are not meant when the reproduceability of measurements is asserted or required, because, thereby, clearly the problem is only shifted to the other measuring device used to control the former. Just as little is meant the historically existing hierarchy of measuring devices, for the measurement results would be relativized to historical indices. If there is something unhistorical in physics (and again, in the self-understanding of physicists, and for good reasons), then it is the hope of also being able to repeat particular measurements, if, say, the historical chain of mutually verifying measuring devices were demolished and we had to begin anew to produce instruments ab ovo. (Let it be recalled that no reference shall be made to units, i.e., that only propositions about ratios relative to *ad hoc* units are intended.) One may thus interpret the claim of reproducibility of measurements to the effect that *the reproducibility of certain properties of measuring instruments, which reproducibility is not dependent on measurement results, is supposed.*

4.3. *Methodology of Measurement.*

After having assumed historical physics in the two-fold claim to be a science of nature, i.e., an experimental science upon the basis of reproducible properties of measuring instruments, as a starting point for a rational reconstruction of physics, we shall now commence with this reconstruction itself. With that, due to the already-mentioned restriction to securing the intersubjectivity of measurement protocol statements, a methodology of measuring is to be developed.

As a basis not further problematicized here, we will (as already in Dingler) assume a standpoint of daily life. It is undisputed that hereby we are concerned with a fiction, since, among other things, our daily lives take place in a highly technicized world which is stamped with a repeated and (at least under certain criteria) successful application of science and consequently is precisely not science-independent. It is, however, also not necessary to give, so to say, a complete description of a non-scientific, life-world standpoint. Rather it suffices to ask at each step of the argument whether or not an argument presented rests on results of a measuring experimental science. By 'life-worldly' is meant then precisely the species of arguments that are science-independent in this sense.

In the life-world, we have at our disposal not only the basic capacities of speech and action, but also craft traditions, as well as an abundance of experiences in the sense of a practice or familiarity with the environment in its natural as well as its cultural, especially technical-artificial properties. It is thus in no way necessary to be 'archaic', to place ourselves back in pre-Gailiean, pre-Aristotelian or prehistoric times or to seek for a reconstruction, say, of craft abilities, techniques and traditions. The rational reconstruction commences with the *qualitative leap from an art of measuring practiced in the crafts to a scientific one*, where the latter is distinguished by certain methodological requirements.

The first step thus shall be made with *linguistic standardization*. The ordinary language already used in the life-world, including technical parts from the area of the crafts, is already at hand but it shall not be simply taken over either as a metalanguage or as an object language. Rather those parts which later will be needed for the terminological construction of the physicist's language are to be defined explicitly once more. As already argued in another context, this definition must begin with the exemplary introduction of predicates which then are to be sharpened further by means of predicate rules. This procedure is delineated in relevant textbooks and in addition is carried out in Chapter III with the example of time-measurement, so that a detailed description is not necessary here.

In particular it is to be pointed out that already the explicit standardisation of the artisan's vocabulary – one should think, say, of certain expressions of the joiner's trade, of the glass-grinder's or of metal work – cannot ensure merely verbally, i.e., by means of illustrations of word usage with the help of other words; it must rest on the acquisition of certain words in connection with corresponding actions, 'operations'. This condition plays an important role insofar as it opens the possibility of securing the validity of elementary propositions by producing the asserted state of affairs itself. Thus, e.g., in the operative foundation of geometry, contact properties of two bodies are used to define further predicates, e.g., the fitting, one upon another, of bodies. For the question of whether, in a given situation, two bodies are touching to be decidable in each case, the pre-geometric predicate 'touch' must be learned together with the action of bringing two bodies into contact. Already on this pre-scientific level propositions will be made possible this way whose validity can be secured by producing the asserted state of affairs 'at any time'. With this, at the same time, the first boundaries for possible objections are also indicated: Obviously the possibility of e.g., grinding two disks against one another finds its limits say in the fact that no grindable materials are present. But no 'theoretical' objections, say from physical theories, that a given action is not possible can be formulated. If an action predicate has been defined, the action is possible *per definitionem* ('in principle', i.e., except for trivial exceptions).

With this, for the most elementary propositions provided for here – they may be called 'basic propositions' corresponding to the first predicates established here, the 'basic predicates' – we have a suggestion for defending them in a dialogue. Even for the domains of the artisan's, the technician's and the physicist's language we shall not strive here for a manner of speech that has degenerated to a permanent monologue, but rather for a humanly communicative one; this means that propositions, i.e., sentences with an assertory character, are characterized by their claim of being able to be upheld by the asserter against questions or objections of a partner in dialogue. For the composition of complex propositions from elementary ones, therefore, we can also use dialogically defined particles (Lorenz, 1968; Lorenzen, 1967; Kambartel, 1979; Gethmann, 1979). In particular, already for basic propositions there is provided here a non-linguistic procedure for defending them, which procedure consists in performing certain testing of manufacturing actions (e.g., bringing two mirrors into contact, moving one body relative to another, etc.).

If one now assumes the availability of a basic vocabulary for a certain area, where the area is marked out, e.g., by the intention to measure lengths or durations or masses, then it is a matter of a second step of *standardizing production and testing* actions by means of rules with the help of the standardized vocabulary. For this purpose *procedural instructions* (*Handlungsanweisungen*) are formulated, i.e., sentences that enjoin given actions. In the instruction to carry out an action *h* (notation: !*h*), an action predicate (e.g., go, grind (disks), apply (a ruler), move (one body)) must occur. Such procedural instructions are given – as a rule as conditional, i.e., "if statement *a* holds, then do *h*!" (notation: $a \rightarrow !h$) – in contexts of action where certain states (properties of devices) are supposed to be produced. In contradistinction to the procedural instruction, an injunction to produce a state *s* (notation: !*s*) shall be called a ***norm*** here. Norms as well as procedural instructions may be designated ***prescriptions***.

The distinction of norms and procedural instructions in the just illustrated sense is of central importance for an operative theory of measuring, even though physicists as a rule are only interested in the properties of measuring instruments (the states *s*), not however in the special procedures (the actions of manufacturing and testing *h*) for producing *s*. The objection made above to Lorenzen's proposal for defining the mass ratio now could be formulated so that, admittedly, a norm !*s* has been proposed, not, however, the procedural instruction (s) $!h_{1,\ldots,n}$, which produce the state *s*. Below, we will still answer objections to protophysics that believe themselves able to rest on the unsatisfiability of protophysical norms. Nevertheless, in addition to norms, the indispensability of procedural instructions also shall be demonstrated positively:

A system of norms, such as e.g., has been established for the operative definition of geometric basic shapes (planar, straight, right-angled, parallel, congruent) yields an explicit definition of special properties of measuring instruments – in the geometric example, rigidity, which is defined as the property of carrying over pairs of points into congruent ones, where this is a property of a measuring stick in transport. Such a norm system is, notwithstanding its prescriptive character, a mere system of sentences and thus does not yet realize the actual operative foundation. One cannot tell, namely, from a system of norms whether there are in fact actions suitable to produce the postulated states. Only the formulation of ***realization procedures*** by means of procedural instructions $!h_{1,\ldots,n}$, for which it then must be shown that they in fact do lead to the required states (i.e.: $s_{1,\ldots,m}$), may count as an operative foundation of a theory of measurement. On the logical relation between elementary instructions and homogeneity principles and on the question how to interprete realiability of homogeneity principles by fulfillability of elementary instructions, cf. Janich, 1984b.

The indispensability of an explicit formulation of realization procedures also exhibits itself in the fact that the norms proposed in protophysics, e.g., homogeneity principles (see below) are propositional schemata. They only become propositions with the insertion of proper names or definite descriptions (Kennzeichnungen) for real bodies and of operatively defined basic predicates. Metaassertions about norms, such as that a norm is unique, only gain significance thusly. Nevertheless let it be expressly noted that the indispensability of realization procedures has nothing to do with the claim that there can be only exactly one realization procedure. Thus, e.g., the three-disk process is one realization possibility for 'smooth' disks, one realization possibility besides other processes that are possible without recourse to already produced geometric shapes, and in particular besides those processes carried out later on highly developed machines. It suffices when there is *at least one realization process* that manages without supposing that the states to be realized have been realized already somewhere else – with a growing technology obviously other processes are resorted to.

As just now mentioned in passing, norms of systems or norms that postulate given properties of measuring instruments must be unique; the *uniqueness* must be proven. What is here to be understood by uniqueness we shall illustrate first by an example: The internal homogeneity principle according to Lorenzen (indistinguishability of points of a surface) is realized, say, by grinding two bodies against each other until they fit together in respectively closed regions. This norm is not unique in the sense that carrying out the realization process two times independently of each other does not guarantee that, for instance, a concave surface of the one pair will fit on top of one of the two surfaces of the other pair. (As a rule different flexures will arise.) It is different with the three-disk process: This process in fact lets one expect that planar disks produced independently from each other will fit on top of each other. (One sees already here that these are not propositions about the bodies worked on, but about the process by which they are worked!) In general, then, norms can be too weak to require the production of 'identical' properties, to require them in the sense that the production processes will be already sufficient reason to make definite statements about the manufactured results of arbitrary repetitions. (Moreover, in another sense, the internal homogeneity principle is also demonstrably unique: Spherical surfaces can always touch each other only in one or in all or in absolutely no points.) The requirement of uniqueness of norms – itself a metanorm – thus implies (beinhalten) so formulating the norms that they are neither too weak to establish logically determined expectations on the relation of different results of the realization procedure [henceforth 'realizata'] satisfying the same norm, nor too strong (in the sense of contradictory, as e.g., the norm of producing a cubical sphere).

After, now, in a first step, having required the explicit definition of the basic vocabulary, which in a second step permits the formulation of procedural instructions for the realization of norms (more accurately: normatively required states), we shall now explain in a third step how a *justified selection of production processes and respective norms* can be made, i.e., how this selection shall be standardized. Again in the sense of the required intersubjectivity of measurement protocol sentences. A realization process for such states (e.g., forms of body surfaces, of motions and of inertial comparisons) will be sought that justifies respectively an anticipation of what relations, formulated in the basic propositions of the theory, hold between independently produced realizata. It already has been pointed out that expectations of this type do not have the character of empirically verifiable predictions because they concern production processes, not however contingent properties of realizata, i.e., of bodies worked on, etc. Thus if in a concrete case a deviation from the expected relations is detected, it can only be inferred from this that the realization process has been faultily executed or that there are external disturbances (presupposing the logical correctness of the deduction of the expectation in question from the description of the realization process). This statement, for its part, is justified by the prescriptive character of the procedural instructions or norms and all propositions logically following from them, to which propositions the distinction true-false has no application. But with that the criterion for what procedural instructions or norms come into question also has been stated already: it must be precisely those norms that are unique that come into question, which *per definitionem* means that they justify definite expectations of realizata.

Obviously, thereby the problem of a norm-governed selection has been shifted to such expectations. What expectations should the assumed realization process justify? With this question one has arrived at the discussion of the ends of measurement in general, of the special ends of measurements of length, time, mass, etc. in particular. A carpenter, the navigator of a ship, a goldsmith, etc. can say very precisely why he measures lengths, times, or masses. Here then it is only a question of which of these life-world ends – possibly modified – should be taken over into physics. With this the discussion flows into the claims illustrated above of carrying on physics as a quantitative experimental science. The criterion may thus be stated wholesalely, one would like *to select those procedural instructions and norms, and set them in play, that uniquely enable a quantitative experimental physics with intersubjectively verifiable measurement protocol sentences.*

Here now a norm presents itself for which it has yet to be clarified in what way it can be satisfied. Dingler and Lorenzen first pointed out this way: ***homogeneities*** must be required because they can be ***uniquely realized*** and also allow the definition of ***criteria for properties of measuring instruments***, which criteria for their part enable an intersubjective verification of measuring results. This is particularly clear in the form of the 'homogeneity principles' proposed by Lorenzen, for which examples have already been given earlier (Cf. p. 57). Homogeneity principles are thus norms that require, e.g., the indistinguishability of points of a surface or of points external to a surface. They require this naturally of a person, and require it of the person who produces such a surface on a body by working it. In the theory of time measurement there is formulated analogously a homogeneity principle for the indistinguishability of individual 'states' that (for a more precise terminological rendering cf. 107) are run through in a motion; with that the production of a uniform motion is required.

What role do homogeneity principles play? More specifically, in addition to standardizing procedural instructions for producing or testing particular properties of instruments, why are they necessary at all? An answer to this question has already been given: To use measuring instruments, it is properties of measuring instruments and not problems of inducing these properties that is important. Homogeneity principles postulate such properties; they are thus, as it were, the complex formulation of production aims; they control on one hand the ***manufacture and improvement process***, and, on the other hand, prescriptively fix the properties of devices to be maintained in the *use* of measuring devices. (Let us already mention now in anticipation, that the process of improving the measuring instrument, e.g., clocks, that actually has come to pass historically can be understood as being governed by a homogeneity principle for the uniformity of the motion of the hands.) To what extent homogeneity principles also govern the improvement of measuring instruments (in the sense of refining them as well as immunizing them against disturbances) will be presented in connection with the following point of view.

With the standardization of the 'basic predicates' from the artisan's language (cf. p. 71), the connection of suitable actions of production with the definition of corresponding words was stressed. Here we now must add that, in addition, suitable actions of selection (e.g., of materials for the manufacture of devices) also must be undertaken. Thus the proposed processes for producing geometric basic shapes ('planar', 'straight', etc.) will only succeed with solid materials. For this, as far as this goes without a methodological circle, we shall establish norms which will then be called *selection norms*. Thus, e.g., in geometry the use of 'hard' materials is required, where 'hard' is defined by the norm of certain transitivities of contact properties. Selection norms differ from production norms insofar as their 'realization process' consists mostly in a mere selection, i.e., cannot generally secure a technical enforceability (Erzwingbarkeit) of the normatively determined states. Where, e.g., there are no suitable materials, the operations for producing measuring devices also cannot be carried out. This dependence, however, concerns only the practice of measuring and makes this practice indeed dependent on particular, i.e., not on arbitrarily, technically enforceable conditions; but it does not make measuring dependent on measurement results (and in this sense 'empirically dependent').

The central significances of *norms* (in distinction to manufacturing instructions) lies in the fact that they achieve the *leap to theory*. The standardization of the production process itself has yet no sort of theory as a consequence; it belongs in the area of official-organizational standardization, as e.g., in West Germany the DIN designations. (DIN: Deutsche Industrie-Norm). Norms such as homogeneity principles, on the contrary, permit an as-if way of speaking, which manner of speech is to be considered as an explicit and logically exact solution to a great number of traditional problems of the exact sciences. If, namely, a norm were employed in argument as an assertion, then this would be materially equivalent to the supposition that there are instruments which satisfy the postulated properties 'purely' or 'ideally'. Therefore let us append to such norms a process of definition in the wider sense that here shall be called *'ideation'*. The process of ideation consists in considering norms without the injunctive part '!', as well as their logical implications. This restriction to a domain of propositional schemata materially implies that real bodies, events, etc. are no longer spoken of, rather the intended properties and production aims are; it will then be indicated terminologically first by replacing basic predicates defined on real bodies (so far as these predicates occur in the norms) by ideative expressions, so-called 'ideators'. We shall then no longer speak of surface segments, edges and corners, but of planes, straight lines and points.

In connection with a methodology of measurement we cannot pursue further here the fact that the process of ideation permits geometric basic terms to be explicitly defined on an operational foundation; this shows the opinion of formal mathematicians that such definitions are impossible and the subsequent evaluation that such attempts at definition are unscientific to be unfounded. Nevertheless let us recall here once more the connection of Hilbert's formalist conception of mathematics with the theory-language concept of the advocates of the Vienna Circle in physics (Cf. p. 3). We discuss in detail in another place the fact that by means of the ideation process a proposal for an exact determination of the status of geometrical theorems becomes possible.

Here we are concerned with the fact that in ideative speech even the *'theoretical' basic concepts of measurement* such as equality of lengths or durations as well as equivalences, and also the rigidity of measuring sticks or the uniformity of motions are explicitly defined. They are then obviously no longer 'theory-laden' in the sense of Hempel, Popper, Feyerabend or Hanson. Rather an ideative definition of such theoretical basic concepts of measurement builds upon a life-world legitimation – i.e., a legitimation that is *per definitionem* independent of the results of any experimental science – of production aims for measuring instruments.

Considered formally, ideation is a peculiar process of terminological definition, different from the various types of definitions mentioned in the literature and most easily comparable to logical abstraction (Cf. Schneider, 1970), where the *tertium comparationis* lies in replacing a previously defined expression (cipher; extended edge) by another (number; straight line) to indicate the restriction to a definite propositional domain. (This restriction is indeed established differently in the two processes: with abstraction it is done by invariance relative to an equivalence relation, with ideation by means of logical derivability from norms.) In symbols an 'ideation principle' can be indicated:

$$(H(r) < A(r)) \Rightarrow (A(r) \Rightarrow A(i)).$$

Here $H(r)$ signifies a homogeneity principle in which operatively defined basic predicates applying to real (thus 'r') bodies, events, etc. occur. $A(r)$ is a proposition logically implied ($<$) by $H(r)$. $A(i)$ is then the same respective proposition in which ideative expressions ('i') occur in the position of the r-expressions, e.g., 'point' in place of 'corner' or 'place', 'line' in place of 'edge', etc. Due to $H(r) < H(r)$ the ideation principle trivially also covers the homogeneity principles themselves. If, in a homogeneity principle $H(r)$, the propositional schemata occurring in it are filled by replacing them with propositions (using basic predicates), then this homogeneity principle becomes a *homogeneity theorem* $H_R(r)$ for a certain realizatum R. Homogeneity principles, whether now formulated by employing expressions that refer ideatively or to realizations, are thus propositional schemata; homogeneity theorems are propositions about definite realizata.

The ideation process permits one not only to refrain from Platonic explanations of idealizations which a physicist favors when he applies mathematics to his experiments. There are no ideal entities proper, and connected with this is the problem of how the world of ideas and the world of real bodies and events harmonize. Even the technical problem of how the use of real numbers can be introduced into the act of measurement – which act for judicious reasons always only arrives at rational-number results – so that finally measurement results can be worked up further with the means of infinitesimal mathematics, is resolvable by the ideation process. The appearance of ideative expressions indicates a precisely determined restriction which may be repeated descriptively as follows: one does not speak of 'real' states of affairs, i.e., of those manufactured on the instruments or measured with them, but rather of the logical consequences which only 'really' would arise if not every realization (in spite of a completely exact determination of the realization aims) were to remain imprecise.

With that, however, the use of ideative terms, which is indispensable in theoretical physics, can be interpreted as an admission of the only incomplete realizability of states of affairs. Norms can, to be sure, indicate in each individual case where a deviation from the postulated conditions is observed, with what aim further improvement is to be carried on. But obviously they assert nothing about the concrete possibilities of such improvements. Attempts at improvement can run up against various limits, e.g., those to be ascertained by means of physical laws (say when the refinement process of instruments approaches molecular dimensions, or when instruments that are 'too large' are supposed to be built with particular requirements). Statements about such limits to realization, however, cannot themselves be defended by the claim made here of the intersubjective verifiability of protocol sentences of measurement, without generally recognizing norms themselves, which are limited in their being able to be carried through. Clearly in the case of determinations of the limit of the realizability of normatively established states, the norms will not be invalidated. (The question of whether, from a possible determination of the limits of realization possibilities, there follow objections to the normative account as a whole will be discussed more closely in Section 2.5 with the help of criticism already presented.)

Here already we shall prevent a possible misunderstanding. The ideation process, which schematically can be very easily extended to other cases (replacement of one n-tuple of expressions by another n-tuple to indicate the restriction to a definite propositional domain), is meaningful as a process for fixing ideative expressions only in connection with unique norms. Ideators presuppose corresponding uniqueness proofs and are thus (in precisely this sense) unique. This distinguishes them from homophonic expressions, say, in formalistic or theory- language- accounts where a comparable claim to unique realizability is not raised.

In addition to this definition–theoretic aspect, uniqueness of ideators has one more feature that is crucial for the theory of measurement: Without this uniqueness, one could not speak of *reproducibility of properties of measuring instruments*. With standardization of the selection of production processes, we mentioned homogenous forms, because for these forms there exist methodologically basic production processes that do not depend on the existence of such forms arising from other processes. Homogeneity thus proves itself to be a discrimination criterion that guarantees that there are production processes in general. The uniqueness of homogeneity principles or of realization processes guarantees, moreover, that production processes, when repeated, in fact permit one to expect the same results. The requirement of reproducibility of properties of measuring instruments divides thus into (at least) two parts: namely, to show that there is a process in general, and to show that this process then always leads to the same results. Only where, in this sense, the reproducibility of properties of measuring instruments has been guaranteed, may one speak of *disturbances* in the case of observed deviations. Measuring instrument norms prescribe the per definitionem undisturbed state; disturbances are then, likewise per definitionem, deviations from these undisturbed states.

This definition of disturbance does justice to the practice that physicists recognize in laboratory work as an obvious truth, namely that disturbances in experimental arrangements appear as deviations from intended states. For the experimenter the artificial production of particular, then, per definitionem, disturbance-free experimental conditions is such a very indispensable component of his endeavors that it is surprising that this input of human achievement to the empirical basis of physics has found no adequate consideration in the affirmative theories of science. Should, in a concrete case, disturbances be observed, then efforts to eliminate them must be undertaken. The process of eliminating disturbances will be called *exhaustion*. This, following Dingler's proposed usage, easily gives rise to misunderstandings which we will prevent as far as possible here.

Exhaustion was spoken of, above all, in connection with the Archimedean procedure of approximating the length of curves (e.g., circles, spirals) by means of polygons. The exhaustion consisted in refining the polygons (raising the number of corners or shortening the polygon sides). This manner of speaking then can be extended to the 'explanation' of natural mechanical phenomena such as e.g., planetary orbits. Ignoring interplanetary 'disturbances' (gravitational interaction of individual planets with each other), the Kepler-orbits can be calculated for each planet first of all from its interaction with the sun. Observable deviations from this (analogously to the deviation of polygon features from curves) are then taken into consideration gradually by means of exhaustion (taking into consideration further planets is analogous to refining polygon features), whereby deviations of observed orbits or orbital times and velocities from the calculated ones diminish.

Although with this we have not explained sufficiently *what* is exhausted – 'nature' in light of available hypothetical calculation results or just these hypotheses – this manner of speaking of disturbances can give one occasion to speak of exhaustion in the case of protophysical norms as well. The normatively determined conditions are enforced increasingly better in (experimental) reality by the gradual elimination of disturbances. The analogue to the observed planetary motions are hereby the ideatively determined states; the exhausting approach takes place in both cases by taking into consideration additional, hypothetically assumed disturbances. Thus we shall also speak fur ther of exhaustion in the case of the elimination of disturbances in measuring devices relative to functional norms, although we shall not thereby conceal the differences with the exhaustion of natural phenomena.

As mentioned previously in passing, with the corresponding selection of homogeneity principles, where the selection is made with this in mind, ideators permit the definition of terms that are crucial for measurement, and that are likewise ideative. In particular, equality (of lengths, durations and masses) will be so defined; on the level of the realization, to equality correspond operations for producing equally large parts (equally long sticks or edges; events lasting equally long; equally heavy or equally inert bodies). To ideatively defined undisturbedness corresponds then, in the realizata, rigidity of measuring sticks, regular operation of clocks as well as homogeneous density (Cf. Janich, 1977a and 1982) in bodies.

It is self-evident that with this we still have obtained no scales. To do this a process of dividing into equal parts must be defined in the domain of length measurement, with the help of ideators, which process then leads to rational length ratios. The definition of 'units' such as the meter does not, indeed, belong here, it does not belong at all to protophysics, because, logically considered, it is only possible with the help of proper names and consequently violates a strict interpretation of the precept of interpersonal verification of measurement protocol sentences. (The reference to units presupposes namely, the availability of a standard of comparison, i.e., is only possible for the persons who have the standard directly at their disposal.) The construction of a scale will no longer be discussed here. It scarcely differs formally from the construction proposed in the affirmative theory of science. It is worth mentioning merely that in the present protophysical account all scales may be reduced to that of length – even those of so-called intensive quantities such as e.g., temperature.

Summary: To secure the intersubjectivity of measurement protocol sentences of an experimental science of physics, the following methodological steps must be taken:

(1) In connection with production actions, a vocabulary on the level of the artisan's language must be standardized.

(2) Therewith certain production actions are to be standardized by procedural instructions, whose selection

(3) is guided by the aim of making properties of measuring instruments reproducible.

Reproducibility presupposes a methodologically first and unique realization process. This uniqueness concerns the aim of production prescribed in the norm, namely a *per definitionem* undisturbed condition or event in a device. Deviations from this uniquely determined condition or event are called disturbances. Uniqueness of norms must be proved.

Restriction to the assertory part of norms as well as to their logical implications is indicated (ideation) by replacing the terms of the artisan's vocabulary by (unique) ideative, theoretical expressions. With the manufacture of instruments, methodological priorities can be detected in the sequence, length – duration – mass. These thus may be called the basic concepts of physics (in the sense of any experimental-scientific physics).

Finally, of epistemological significance is the fact that in the opposition of normative definitions of undisturbedness based upon the foundation of operations to 'external' disturbances that appear as deviations therefrom, the distinction 'non-empirical' – 'empirical' acquires a precise sense unburdened by the interpretation problems of Kantian *a priori*: The result of a measurement is *empirical*. The intersubjectivity of measurement results, however, is based upon the non-empirical undisturbedness of measuring instruments, i.e., upon an undisturbedness not depending on measuring results but rather one normatively secured by man.

But with this the statement that there are basic concepts of physics also loses its closeness to fundamentalistic or ontological views. The basic concepts of length and duration are not fundamental for physics because space and time are 'basic categories of human cognition', rather they are fundamental only relatively to the designation of physics as an experimental science; if experiments are to be done on instruments with reproducible properties, then above all spatial forms first must always be produced on instruments. And than should, events be compared with a reproducible standard then standard events, i.e., operations of clocks, must be produced. The labelling of length, duration and mass as physical basic concepts owes itself, thus, to the characterization of physics as an experimental science and in textbook presentations is buried under the restructuring of physical theories that has been unloosed by experimental praxis.

5. ON THE CRITICISM OF PROTOPHYSICS

A series of authors have criticized the first edition of this book as well as other works on protophysics. The criticism and a reply to it is handy and easily surveyed in other places. (Cf. Böhme, 1976; Kamlah, 1981; Pfarr, 1981; Tetens, 1984; Janich, 1984; Pfarr and Janich, 1984.) Moreover this criticism is taken into consideration in the present second edition. Nevertheless it is not superfluous to once more here go into some points of this criticism, since it has spread, seized upon opinions stemming from the theory of science called 'affirmative' in this book and marched argumentatively into the field against protophysics. It is, namely, to be expected that many readers not familiar with this criticism will come to similar objections, so that we will give here a thematic repetition of some points of the criticism proffered as well as a short reply:

1. *Objection*: Protophysics formulates as the norm system for measurement of length a Euclidean geometry, as the norm system of measurement of time a Galilean kinematics, and as the norm system for measurement of mass a classical dynamics. It thus stands in contradiction to the results of modern relativistic physics and is therefore to be rejected.[16]

This objection can (a) be so meant that particular norms of protophysics (e.g., the external homogeneity principle of Lorenzen) contradict particular theorems of physics. This is, however, not possible for the trivial reason that norms generally cannot come into logical contradiction with propositions. Norms prescribe something, propositions assert something. Norms are either justifiable or not; propositions are true, false or undecidable, norms are none of these. In this sense, then, the objection certainly can not be maintained.

More frequently then, the objection is so advanced (b) that modern physics has already provided knowledge that the protophysical norms cannot be followed. Protophysics indeed does not stand in contradiction to physics, but as a system of norms that cannot be followed it is vacuous and irrelevant for physics.

In fact, protophysics could not make good its claim to be a rational reconstruction or foundation of physics if it did rest on unfollowable norms. Therefore we must clarify whether the protophysical norms are known to be unsatisfiable.

Almost all attempts to prove this unsatisfiability rest on physical knowledge, i.e., argue e.g., against chronometry with special relativity theory. If this objection is made with the claims to intersubjective validity of the physical theory cited, it itself presupposes what it attacks. It is therefore methodologically circular and thus irrelevant. If, however, it does not insist on this intersubjective validity of the physical theories cited, it is not clear how it is justified at all as an objection. The most sharp minded form of this objection consequently argues that protophysics cannot preclude the possibility that nature is as e.g., the relativity theories assert;[17] the validity of relativistic physics thereby is not asserted. This objection may possibly touch older views of Dingler; it does not touch protophysics: for protophysics asserts nothing about nature, but rather about production processes as suitable means for particular ends; it also makes no statements about nature in the sense that it would assert that certain states of affairs could not be real. Protophysical norms exclude no empirical states of affairs at all.

Here also belongs the empiricistic objection (c) that in the end protophysics presupposes the followability of its procedural instructions and norms. Whether these presuppositions are satisfied can only be known from experience, for it is thereby a question of statements about reality. This is based on the conception that protophysics depends on the sentence 'Procedural instructions (or norms) are satisfiable.' This sentence does not occur in the construction of the protophysical parts of geometry, chronometry and hylometry themselves, i.e., must not itself be grounded.

In contrast protophysics can claim for itself that (1) through the complete operative definition of the basic vocabulary, through freedom from contradiction and through uniqueness, the step-wise followability of the production instructions is secured; that (2) through recourse to the historical achievements of a non-scientific artisan's trade, the followability of the proposed procedural instructions is an historical fact; that (3) protophysical norms systems can be considered as adequate reconstructions of the scientific art of measuring; and that (4) up to now no single rival theory is known which, with consideration of pragmatic or action-theoretical aspects, enables non-circularly a historical and systematic understanding of the scientific art of measuring as it is in fact practiced.

2. *Objection*: Measuring instruments are, as parts of the physical universe, subject to the same laws as all bodies, occurrences, etc. Therefore, independently from our intentions in producing and using instruments, properties of measuring instruments must be consistent with natural laws.

That measuring instruments are part of the universe and, besides artificially produced properties, also have natural ones will certainly be disputed by no one. Where the difficulties already mentioned under 1. do not appear, i.e., that objections from physics are either circular or not legitimate, there still remain open at least the following points of this objection:

Where in general properties of measuring devices are to be exactly empirically knowable, they can be so only either with the help of other measuring instruments (which either leads to an infinite regress or to a dogmatic interruption of further questioning) or upon the foundation of norms for measuring instruments. A restriction to empirical properties of measuring devices, and therewith the treatment of artifacts as natural objects, is always a grievous error; it presupposes blindness to the history of science and current research practice and can contribute nothing either to the theory of measurement or to empirical theories.

A *de post facto* explanation of properties of measuring instruments within the scope of empirical theories is indeed methodologically possible, but remains obviously after the fact and can achieve no foundation of a theory of measurement or of the practice of measurement.

Frequently the above-cited objection is made in the form that the followability of protophysical norms for the order of magnitude of our instruments is granted; in respect to relativistic or quantum physics, however, it is to be considered as approximate, and in microphysical and cosmic dimensions a realization would not succeed. To this is to be answered (1) that in fact we are concerned only with instruments and not with creations of fancy such as mirrors of cosmic dimensions or similar vagaries, (2) that, *de facto* and in accord with protophysical theory, deviations in measuring instruments from normatively determined properties can always appear only 'technically', i.e., not as deviations from any nature structured by laws, but as deviations from a standard of precision defined for a technical application. The production of measuring instruments is thus either possible or not possible. The assertion that they would be approximately possible is nonsense.

3. *Objection*: Protophysics is anachronistic. It orients itself to the paradigm of classical physics.

In fact, protophysics attempts to ground the measurement of length, duration and mass operatively. It does not begin, thus, e.g., with the quantum-mechanical process of measurement. This must, however, appear as an anachronism only where there are available no other criteria than adhering to, in terms of the theory of science, the respectively most recent physics textbooks and assuming unquestiongly the self-understanding of physicists offered there. As already presented in detail, protophysics tries to make good the claim to intersubjectivity of measurement protocol sentences. For this, that the quantum-mechanical measurement process should have priority over the classical, can only be argued with physical theories which themselves depend on measurements. Starting with the quantum-physical measurement process would thus be a sure means of barricading the non-circular entrance to the intersubjectivity of an experimental physics. An analogous remark holds if beginning with a modern formulation of electrodynamic theories were demanded. Positively stated, efforts at foundation may not begin on a level on which results of a measuring physics are already presupposed.

From this, indeed, it does not yet follow that protophysics is oriented to the paradigm of classical physics (with the suggestion that with orientation to a modern paradigm there also would result a protophysics leading to modern theories). Obviously protophysics has developed subsequently to classical physics, but this also applies to the most recent accounts, e.g., of quantum field theory. Protophysics nowhere depends on historical authorities and also does not set itself the goal of reconstructing classical physics 'for the first time', parallel to the history of physics. It also does not attempt to follow unquestionably a classical cognitive ideal (e.g., Euclidicity and Galileoinvariance). The starting point is rather the question of with what means we would have to begin physics today, if we had at our disposal no results of any kind of modern physics, in order, with these means, to pursue precisely described intentions. In this sense, then, protophysics shall permit a critical judgment of physics: can our present-day physics be understood as an adequate means for attaining our present-day intentions? Or is it merely an extension, in the meantime, of old intentions that are no longer understood, and possibly intentions in accord with foreign ends, e.g., those produced by institutionalization?

4. *Objection*: Protophysics makes too high a claim.

Here those critics shall not be enumerated who attribute to protophysics a claim which it does not make, and hang thereon the proof that it cannot make this claim. What then remains for objections follows definite thought patterns, as e.g., that knowledge of the world must be able to founder on experience. Above all, in the critical rationalism of Popper and his followers, openness to a revision of doctrines at any time, is recommended with empiricistic pathos as a virtue of modesty. It is more appropriate and modest to simply reckon with the foundering of every cognitive effort, i.e., without the specification 'founder on experience'. Obviously protophysical theories are also exposed to this risk of foundering; only empiricist attempts to prove that they have already foundered are not defensible. Even the revisability of protophysics is not disputed by its advocates. Grounds for revisability extend from proof of logical or methodological mistakes to the change of ends which are now attributed to physics, now to protophysics. To derive from the fact occasionally noted by protophysicists that, at present, there is no rival theory to protophysics with a comparable cognitive interest, the claim that protophysics is the only way to found physics is logically false. Finally, let us add that in protophysics or in the methodology supporting it, no new goals, either of a physical or methodological sort are set, and that in contrast to the affirmative theory of science the endeavor of protophysics follows upon the old claim, made at the beginning of the Vienna Circle, to define the scientific nature of physics.

CHAPTER III

CHRONOMETRY

1. WHAT PURPOSE SHALL TIME-MEASUREMENT SERVE?

After the critique in Chapter I of common views of concept formation in physics and then specially of the views of time-measurement, we presented in the methodological discussion in Chapter II in what way the basic concepts of physics can be defined. The principles developed there concerning the construction of a physical terminology, more specifically, concerning the non-circular beginning of such a construction, are to be borne in mind if now a theory of time-measurement is to be established. To be sure, these general methodological principles alone do not suffice for this; they merely relate to guiding principles for the procedure involved in the definition of a physical basic concept; however, they say nothing about where to begin in the case of a specific basic concept such as e.g., that of duration, or towards what goal the terminological construction is to be directed. Here, therefore, the construction of a theory of time-measurement will begin by considering what expectations should be connected with the practice of time-measurement; or, to follow the terminology of Chapter II, what purpose time measurement should serve.

In discussing the question, what purpose time-measurement is supposed to fulfill, naturally, in a broad sphere a preliminary understanding enters, an understanding which has arisen in a very complex manner. It is easy here at the outset, to distinguish between expectations made of clocks within the context of the experimental sciences and those which we have in employing clocks in daily life. Nevertheless this distinction has its difficulties, because our commerce with clocks in daily life and, consequently, also our preliminary understanding of time-measurement which does not directly concern the sciences are often influenced by technical and theoretical developments from physics. Thus today not only are quartz wristwatches in use, which, in terms of running accuracy, were hardly available to physicists fifty years ago, but also electric 'clocks' in the household, which for the most part are only indicators of the voltage oscillations of alternating current and clocks with digital dials also belong to the devices used unreflectively in daily life. Consequently, beliefs such as those that electric clocks measure time more accurately than mechanical clocks, or that the correction of incorrectly running clocks consists in synchronizing them with official clocks etc., arise just as easily as does the conception of a single time (the official time of radio stations within geographical time zones). Already children acquire words referring to time with the help of clock-faces or clocks. The difficulty of time-measurement, namely artificially producing a process and maintaining it as free from disturbances as possible in order to measure time by it, and relating such artificial time-measurement to natural time-measurement, say, with the aid of sun dials, is not recognized in the everyday understanding of time-measurement.

However, even physicists have no reflective preliminary understanding of the purposes of time-measurement. To the theorist time is a real, and for about 70 years a place-dependent, parameter. When he himself measures time, he does so also only as the possessor of a wristwatch and in the context of everyday tasks of organization. However, even the experimental physicist who works with accurately running clocks has no reflective preliminary understanding of time-measurement. As a rule it is he who buys the laboratory clocks. He depends on the manufacturer's data about the guaranteed running accuracy, and should suspicion of deviations in their operation arise he will, if he does not have the clock repaired by a specialist, check his clock with the help of physical experiments and with reference to physical laws. Whether these laws themselves must be checked with clocks of a definite sort is of no consequence to him. Thus the experimental physicist uses clocks in such a way that whatever does not produce measurement results contradicting recognized theories is taken to be a good clock; or expressed pointedly: where the regularity of the operation of a specific clock stands in question, the physicist always decides in favor of the clock which corroborates those theories that the physicist otherwise believes.

It is no objection to the success of clock technology and time-measurement that, not only in daily life but also in science, clocks are used unreflectively as a rule, i.e., that to the user of the clock the reasons why he is successful with specific clocks are not known. However, what thereby counts as success is discussed neither in physics nor in the theory of science within the framework of theoretical reflections on time-measurement; the criteria of success lie outside theories of time-measurement.

Should one wish to seek the purpose of time-measurement by analyzing the conditions under which the user of clocks in daily life as well as in physics *de facto* considers his time-measurement to be successful or to have succeeded, then one would thereby only be attempting to impute purposes after the fact to current practice. The result would have to be historically foreshortened, i.e., the purposes which in the history of watch-making led to the clocks which we use today could not be recognized in this way. Against this historical foreshortening it may be objected not only that technical or scientific accomplishments cannot be understood as merely the result of specific efforts, namely efforts directed to specific goals, but also possibilities of judging the practice of time-measurement and its theories are lost if a practice, as it is now to be encountered, is investigated in regards to what it accomplishes, and if these achievements are then taken after the fact to be purposes found by means of analysis. It could, namely, well be the case that the purposes imputed to ordinary time-measurement after the fact are other than those originally pursued by the designers and builders of clocks. Under these circumstances one could speak of a successful watch-maker's art only with restrictions. If therewith time-measurement techniques in their success are unhistorically referred to currently recognized theories, then, by means of the imputation of the validity of these theories, to answer the question of whether these time-measurement techniques are suitable means for testing theories becomes impossible. In other words, it is not a question here of finding a theory for a sphere of physical measurement which presents itself in the form of physicists' habits of action, i.e., of assigning a theory to the current practice of time-measurement in physics. Rather the foundational attempts employed subsequently in this book are justified precisely by the fact that the expediency of a practice of time-measurement based on an historical process of development is investigated critically. Consequently if the use of clocks by physicists and the functional criteria for clocks accepted by physicists no longer remains the given starting point of the consideration, then naturally the purpose of time-measurement can also no longer be found by beginning with the question, what do our modern clocks achieve? Rather the first question must now be: What should our modern clocks achieve? And to answer this question no physical theorems may be used whose acceptance can be obtained only with the help of clocks. Otherwise a foundational circle would present itself.

To further clarify this important aspect of a foundation of chronometry let us characterize the intention of establishing a chronometry as follows: We are not concerned with an analysis of the present-day art of time-measurement, an analysis whose result would then be called the theory of time-measurement. However, we are also not concerned with giving a mere description of the history of the art of time-measurement. Rather what is sought is a normative basis, i.e., a norm or system of norms which prescribe the production of specific states of affairs, such as the ways clocks function, as well as instructions for action and speech which enable a satisfaction of the norms.

In this way, admittedly, we question what lies behind the practice of using clocks, a practice accepted as proper by (nearly) everyone – something frequently evaluated, particularly by theorists of science schooled in positivism, as an unintelligible artificiality. To be sure, neither an offence against the primacy of practice lies in this questioning, nor is such questioning methodologically dispensable; since the practice of chronometry investigated here is, as pointed out above, no longer a direct, elementary everyday practice, i.e., it is no longer independent of scientific results, even results based on time-measurement. Theory-dependent knowledge has already entered into the construction of all modern clocks.

That in the case of chronometry we must first descend from an already attained complexity of practice, thus illuminates the character of its methodological foundation. The principles, i.e., the beginnings are of interest here; introduction of first words for an exact speech about time and examination of the possibility of building clocks without having already available knowledge of the lawfulness of 'natural' processes can be such a foundation. To be sure, for this we need not ignore the successful use of clocks for more than two thousand years, nor from the outset need a non-circular chronometry pursue the goal of being taken into consideration in the production program of a modern watch factory. Moreover the non-circular execution of the first steps shall provide a contribution to the theories of the physicist in which chronometry, above all since the discovery of special relativity theory, plays a significant role; this contribution is not so much for the purpose of closing a gap in the construction of some theory or other assumed to be valid – for aesthetic reasons, as it were – as rather for gaining criteria for distinguishing 'good' and 'bad' theories.

At this point further search for a 'foundation' of the normative basis itself may be discontinued. Since every normative basis occurs with no claim to propositional truth, but rather only furnishes the possibility of arguing for or against the truth of propositions, *qua* assertion it does not need a foundation. Such an attempt, e.g., proving the need of criteria for distinguishing good and bad theories, would instead depend on giving a moral argument for normative speech actions. With this, viewed methodologically, one would enter into a new domain, namely insofar as natural science and the discussion of its methodological principles is pursued, above all, today precisely with a conscious down-play of moral argument. That this methodologically justified restriction has become a principle which has acquired fundamental significance for the self-understanding of most exact scientists may be regrettable; the attempt to fight against this, however, will be restricted here to demonstrating the possibility of furnishing criteria for distinguishing 'good' and 'bad' theories.

After these preliminary remarks it now becomes a question, from a preliminary understanding, of exactly defining goals which should be supposed for a chronometry. By using the simplest possible words of ordinary language we may show what temporal propositions occur in everyday language and what the needs are which require an exact definition of speech about time, in order finally to require support of this speech by artificial aids such as clocks.

Without claiming completeness we may assert that the most frequent temporal propositions about changes are those that concern *the ordering of events* in respect to earlier or later, or again those that concern the *duration* of events. These propositions shall be called *pure temporal propositions*. By *mixed temporal propositions*, then, will be understood propositions that treat of velocities and forms of movement. They are (even in the clockless and prescientific domain) reducible to pure ones and can be considered as an extension of the domain of pure temporal propositions by including spatial propositions. To neither of these two groups belong those say which make use of the distinction past-present-future. It is in no way settled that this distinction will not be able to become relevant for exact science. However, as long as no direct need has been established for including this distinction in the foundational discussion of physics, and since the distinction of pure and mixed temporal propositions does not depend on the modes of time, the modes of time will not be considered further here.

Thus, in first approximation, the task of chronometry may be declared to be that of exactly enabling pure and mixed temporal propositions by constructing a terminology and formulating norms for the functioning of devices to exactly determine time.

For this purpose let us briefly consider clock-free temporal propositions with the help of ordinary language and its peculiarities. To speak of the ordering of 'outer', i.e., observable, events is possible in at least two ways: Either one makes use of spatial metaphors with the help of the picture of a punctiform present travelling uniformly along a time line, dividing the line into past and furture; or one refers to a direct or imagined immediate experience of the series with the help of words such as 'now', 'then', 'before', etc.

From the domain of spatial metaphors arises the constantly advanced (and, strictly speaking, unfounded) opinion that time is 'one-dimensional'. This peculiarity coincides with the asymmetry between the possibilities of spatially ordering bodies and the possibilities of temporally ordering events. Two bodies cannot occupy the same place simultaneously, while, however, two events can occupy the same time 'at the same place', i.e., occur at the same place at the same time.

Notwithstanding such familiar speech customs originating in ordinary language, it is very well possible to define order-relations for events which are analogous to the spatial relations of bodies. One could imagine perhaps non-overlapping events temporally represented by pebble-flints which are arranged in a straight row corresponding to their temporal succession registered in immediate experience. By laying stones next to these to represent simultaneous events, a two-dimensional representation could be obtained. Here it may be objected that with pebbles, i.e., bodies, one already knows that from rows a layer can be formed, from layers a pile, but from piles only piles can again be formed. With events one would have no comparable knowledge. The lack of such knowledge indeed does not lie in events and bodies, but in the distinctions that we make in speaking. These distinctions, for their part, are accessible to historical explanations if need be. For millenia all organizational needs were satisfied just by carrying out length, surface and spatial determinations and fixing order-relations for the objects of barter and commerce. Only at a much more recent and relatively widely developed state may more accurate determinations of time have supervened, say for travel times or a calendary determination of time. Presumably, arguments may be found for the view that even the needs of a modern industrial society are not so differentiated as to establish the practicability of talk about a multi-dimensional order-structure of events. What should be presented in these remarks on the spatial-metaphorical talk about the ordering of events is the mere historical character of distinctions which exhibit themselves in familiar phrases of ordinary language. To speak thusly of a compulsion 'arising from the objects' and to say nothing else about their order, on the other hand, cannot be speech. Where a critical reconstruction of customary order-relations does not succeed or again where it appears as inexpedient, other relations are to be introduced and used.

Non-metaphorical talk of temporal order distinguishes simultaneous and non-simultaneous events. Non-simultaneous events are ordered according to earlier and later. (Finer representations, in which e.g., a temporal overlap of events can be expressed, produce nothing new in principle and can consequently be passed over here.) The question, then, is: How are 'earlier' and 'later' introduced into ordinary language and from this introduction does an argument result for incorporating this distinction into a chronometry or, on the other hand, for even forbidding its incorporation?

So far as 'earlier' and 'later' designate a temporal relation to the speech situation, i.e., so far as they are used – in the terminology of logic – as 'indicators', their use can, at the outset, be prohibited for texts of exact science, since the theorems of exact theories must be situationally invariant. As for what concerns the predicative use of 'earlier' and 'later', the exemplary introduction may follow from two groups of examples: (1) In a conversational situation two events (at one place) are observed; they bear proper names or distinguishing attributes, and then to one of the two the predicate 'earlier than' the other event is applied. (2) Instead of a reference to immediately experienced events, the series of events is pointed to with the help of 'indices', i.e., the predicate 'earlier than' is introduced by indices. Thus, say, the narrower growth rings of a tree trunk came into being earlier than the broader ones. For this second mode of introduction and use, knowledge is required which already presupposes knowledge of the distinction of immediately experienced events in terms of earlier or later. This knowledge can itself, in special cases, reach the range of a scientific discipline, such as e.g., with the geological determination of the age of rock strata. Since, however, with the foundation of chronometry, we should and can also first provide the presuppositions for gaining exact knowledge about the assessment of 'indices' in fixing the succession of physical states, the second type of examples must remain out of consideration. It is unavoidable to examplarily introduce 'earlier' and 'later' via immediately experienced events. Since, however, every dependence on accidental experiences must be avoided and since furthermore the demand for verifiability of basic propositions has been laid down, procedural instructions for the production of suitable examples must be given which lend themselves to the exemplary definition of 'earlier' and 'later'. Presumably this task should contain to difficulties in principle, particularly if recourse is permitted to already provided spatial order-relations. As is well-known, a problem first arises when (presupposing that 'earlier' and 'later' have been introduced as basic predicates) the succession of spatially distant events is to be determined. Then some sort of decision procedure is necessary. Except for such procedures which already themselves presuppose a chronometry (e.g., use of light signals, propositions about whose propagation velocity themselves are only gained with the aid of clocks or with the help of a transport of clocks) no procedures are known up to now which, except for a mere theoretical significance for physics, would actually be utilizable.

To be sure, after defining 'rigid body', rigid rods can be imagined at two distant places, which, by means of underlayed transport, permit at one place a decision about the order of succession of two distant events; but it can already by defended pre-chronometrically that this logically correct procedure is practically unworkable.

Consequently the following holds for the use of the predicates 'earlier' and 'later' in chronometry: No objections arise against their introduction at a place at which procedural instructions for the realization of events can be followed. Nevertheless a decision procedure for spatially distant events cannot be given so elementarily that 'earlier' and 'later' may be introduced as basic predicates. They have their place at a later point of the methodological construction.

Besides comparisons of events in regard to their order, comparisons in regard to their duration were also ranked among the group of pure temporal propositions. Strictly speaking, everyday language recognizes (again free of clocks) only propositions about the relations of durations. If one merely asserts of an event that it lasts long, then in defending this proposition we must have recourse to a comparison event. In everyday practice such comparisons of events present no problem at the outset. Questions about the individual steps which are thereby run through can indeed show that at least one group of time comparisons cannot manage without definite conventions. Propositions of the type 'The holiday lasted as long as the period of good weather', which assert the equal duration of two simultaneous events, are unproblematic. Much more frequently, propositions about duration follow the purpose of giving a conversational partner the possibility of placing a specific event itself into relation to arbitrary events (say, events from his memory) i.e., in particular to compare non-simultaneous events in regard to their duration. The attempt to apply an intersubjective 'measure' in this sense requires some sort of determination of standard events, a determination fixed in the speech community. The change of day and night, the moon phases and seasonal changes have offered themselves since time immemorial for this purpose.

For non-scientific practice the selection of standard events is sufficiently justified by empirically proven expediency or also by traditional reasons. Nevertheless within the scope of an exact science other justifications are demanded, about which we will yet speak. To be sure, historically, the unfounded selection of standard events has extended itself into physics. There are no non-circular foundations for the reference of temporal determinations to the earth rotation, pendulum movements or light diffusion.

With the group of pure temporal propositions was contrasted the group of mixed ones. Here we must first exactly define the criterion which distinguishes mixed temporal propositions from pure temporal propositions. While comparisons of events with regard to order and duration require no sort of restriction to specific events, in the talk about forms of motion and of velocities restriction to natural-scientific and technical question formulations already shows itself; the reduction of talk about 'nature' to what is geometrizable may already be recognized. Here the reduction of arbitrary changes to elementary spatial or material models becomes obvious, which reduction (disregarding the possibility of analytically true sentences) is frequently considered as the sole basis upon which rigorously verifiable propositions would be possible. To be sure, forms of motion and velocities may also be spoken of without reference to local motions of bodies ('The forest changed color slowly at first, then almost suddenly.'); but even changes such as changes of color, temperature, etc. are finally reduced to measurements of length in physics or at least represented spatially.

In regard to mixed temporal propositions, we shall expect of a chronometry that assertions about forms of motion and velocities will be formulable and operatively verifiable. Here forms of motion offer a certain difficulty, since to define them analogous to everyday speech appears, above all, possible by means of one-place predicates; however the requirement of the testability of all propositions within the framework of an experimental science seems to demand the use of multi-place predicates. What (in the following, always with restriction to local movements of bodies) is to be understood by 'form of motion' is determined by definition: In that one asserts or denies (exemplarily introduced) predicates such as 'jerky' (*ruckartig*), 'regular', 'accelerated', etc. of given motions, one forms propositions about 'forms of motion'. The predicates which distinguish forms of motion are one-place predicates. If now the requirement of operative testability of propositions about forms of motion is raised and if one excludes as a testing procedure the simple appeal to a dialogue partner to once more consider the examples of typical forms of motion (this procedure would amount to a continuous extension of the domain of examples for the introduction of predicates and could only lead to asserting new singular sentences) then a special procedure for defending propositions about forms of motion must be given. Since, due to the requirement of situational invariance, the protophysical decision procedure always amounts to comparing an object in question with a reproducible standard, it may already be seen that the predicates accessible to such a concrete dialogue must be two-place predicates.

Thus, consequently, we may first only assert that a motion (e.g., of a body on a trajectory) is uniform, jerky, accelerated, etc. in reference to another simultaneously occurring motion. This however does not exclude that an additional definition of one-place-predicates for forms of motions can be given on the basis of relational definitions and without any reference to an 'absolute concept' of time or a concept of absolute time. Some critics of protophysics did not understand this two-step-procedure and quoted E. Mach against protophysics: "A motion can be uniform relative to an other. The question however whether a motion is uniform as such does not make any sense." (Cf. Pfarr 1976, p. 149, and Mittelstaedt 1976, p. 17 f.)

The case is similar with propositions about velocities. The predicates customary in everyday speech such as 'fast' and 'slow' are likewise implicitly two-place predicates. The context allows us to recognize in reference to which motion the motion talked about is fast or slow. Here it is evident that determinations of velocities and of forms of motion in everyday speech are independent of each other, i.e., motions of every form can have different velocities and conversely. The difficulty offered by a comparison of the velocities of non-simultaneous movements is analogous to that encountered with judgment of forms of motion: Since the human 'time sense' cannot meet the demand for certainty made by protophysics as well as by physics, a reproducible standard must be provided.

The requirement of reproducibility of velocities and forms of motion by machines results in a further aspect for the expectations made of a chronometry which goes beyond the achievements of everyday speech in respect to temporal propositions: subsequent to the acquisition of temporal expressions through a speaker, pure temporal propositions may be distinguished from mixed ones. However this division into two classes implies no methodological dependence of all or some propositions of the one class on those of the other. I.e., one can state no methodological reasons why, at first, propositions of one of the two classes would have to be mastered in order to be able to meaningfully use or learn those of the other. This is supported not only by the circumstance that in a child's natural learning of his mother tongue, expressions for pure as well as for mixed temporal propositions occur equally without a succession of learning steps being able to be distinguished in general.

Also within the scope of scientific endeavors it presents neither a logical nor methodological problem to speak about events, to define terms and finally to form temporal propositions about events without using geometry. There also exist no methodological objections to the possibility of systematically ordering such propositions, formalizing them and perhaps constructing an axiomatic theory. Such a formal theory might also be called a 'theory of time'. However it is a fundamental misunderstanding that can be observed among physicists as well as among theorists of science to think that such merely verbal or merely formal theories could be a foundation for a measuring physics of whatever nature. This is because an empirical physics is not possible without a systematic production of instruments. Only where those steps also enter into the theory, steps which assure a reproduction of standard events or standard motions, can one speak of a theoretical basis for a measuring physics. Beyond this, the merely formal approach overlooks the fact that neglecting the circumstance that without experimental practice no physics can be carried on not only fails the claim of physical propositions to validity, but, even more, fails the theoretical as well as practical justification of the endeavor 'physics'.

Thus the numerous approaches to formal theories of time, which with aids such as set theory, group theory, topology, etc. have been elaborated into axiomatic constructions, remain unproductive for physics, in spite of their logical correctness. It may namely already be established prechronometrically that a geometry-free talk about events, i.e., a theory restricted to pure temporal propositions, does not suffice for more than merely formal operations; thus it can have no relevance particularly for actions which must be carried out by the experimenter in the fulfillment of time-measurement: The claim of physical propositions to validity at every place and every time excludes the choice of accidentally observed events for standard events. The reoccurrence of equal standard events as well as a defense strategy for the assertion that an event is the repetition of an earlier one first attains, then, practical as well as theoretical availability when it becomes a question of artificial events whose equality is enforced.

In summary, without having recourse to theorems of theories, which for their part can have no validity without time measurement, we may state the following requirements – where only a preliminary understanding is drawn upon, an understanding which is independent from scientific results and is visible in ordinary language or results from planned usage-contexts of clocks:

Firstly, a terminology must be furnished which permits temporal propositions at the lowest state, i.e., permits temporal propositions to be formulated without reference to general standard events or motions. In view of the goal of founding a technique of time-measurement and thus consequently of also comprehending the artificial production of standard events or motions with the help of instruments, this terminology must already begin with mixed temporal propositions, i.e., with predicates for the comparison of velocities or motion forms. It may also be understood that these propositions must be about motions of bodies. Thus, consequently, the construction of a chronometric terminology has to begin with the comparison of simultaneous motions of bodies. It may also be asserted from the outset that, consequently, propositions about the duration of (simultaneous) periods of motion are already possible, such as e.g., that they last equally long, or that one motion lasts longer than the other; moreover simultaneous motion of bodies can be compared in regard to their respective momentary velocity as well as in regard to their (relative) forms of motion in a type of integrating proposition about the momentary velocities.

An extension regarding the tasks of time-measurement in the narrower sense is necessary for comparing non-simultaneous motions. Here a selection of standard motions must be made possible for which a situationally invariant repeatability can be secured. This last-mentioned viewpoint, namely the requirement of the (artificial and thus optimally independent from individual situations) reproducibility of standard motions for time-comparison, allows us, from the outset and without knowledge of physical theories, to exactly define a series of further characteristics of the sought-for practice and theory of time-measurement. This runs completely counter to the most widely prevailing opinions of physicists and theorists of science, for whom the art of measurement is a part of physics and completely dependent on empirical theories.

Prior to any empirical physics, we may establish the goal of enabling a *situationally invariant comparison of time*. Naturally, for this, an understanding must already be present of what, in general, a time-comparison (of non-simultaneous events or motions) is. For this the human time sense comes into consideration as much as reference to natural events. However, situational invariance is a requirement which is then raised in addition to a spontaneously arisen ability for time-comparison in daily life. Classically, in the place of situational invariance, one frequently spoke of the independence of time-comparison from place and time. To be sure the world view of classical mechanics already underlies this conception. Situational invariance must rather be defined as independence from the experimenter; and only thereafter, when a measuring physics has already come into play, can this independence from the experimenter be recognized as an independence from place and time and, moreover, from still other factors.

Thereby is shown that from the requirement of situational invariance of time comparisons we can already derive definite restrictions for the appropriate standard movements. Yes, beyond that, even before that, something about the form of the sought-for standard movement may be stipulated, since the independence of chronometry from time as a special case of situational invariance concerns not only the rough date (we would perhaps say in everyday speech: independence from the calendar date of the day on which the time-measurement in question is given). In the ordering of the magnitudes of the motions or parts of motions considered in the time-comparison, independence from time must also be guaranteed. By this is meant that not only must it be insignificant as to when a time-comparison is carried out, but that also the standard movement itself may not be the cause for differences being ascertained in the movement compared with it – differences which do not show themselves or show themselves differently as soon as the standard movement is temporally displaced insignificantly against the observed movement. In other words, the standard movement must be *homogeneous*.

The homogeneity of the standard movement, as a consequence of the requirement of situational invariance of time-comparison, might be easily grasped if on one hand a parameter representation of motions of bodies were already available and if on the other hand such parameter representations were already applicable to non-simultaneous movements. Then the requirement of homogeneity would be first equal to the requirement that with a time t, time or movement comparisons undertaken must exhibit the same results as with a time $(t + \Delta t)$, where Δt is an arbitrarily chosen temporal shift of the standard motion (more precisely: of a definite part of the standard motion).

Beyond this the requirement of situational invariance may also be exactly defined accordingly, namely that it should be the task of measurement-dependent time-comparison to obtain propositions invariant in respect to magnitude. We need not discuss whether, prior to any science, reasons may be given for why an empirical science based on comparisons with the aid of instruments should be invariant with respect to magnitude, since in any case such reasons may be stated with certainty prior to any time-measurement. For this purpose one can have recourse to geometry and thereby explain what 'invariant with respect to magnitude' means. Propositions such as e.g., the Pythagorean theorem hold invariantly with respect to magnitude. No reference is made in them to a standard magnitude. Where magnitudes such as lengths of segments, surfaces and such like are compared with one another, only ratios of magnitude can be obtained. Here the undoubtedly successful use of units in physics shifts attention to the fact that a magnitude-invariant or unit-invariant geometry must first come into play for reference to units to be methodologically meaningful.

Transferred from geometry to chronometry, this means that the results of time and motion comparisons must also be independent from the velocity of the standard motion. A modern physicist would express this state of asserting the invariance of physical laws relative to the transformation of time

$$\tau = at + \Delta t$$

i.e., every physical law which holds according to the time-measurement t, also holds according to time-measurement τ.

Such an assertion is naturally irrelevant for the reflection employed here on what requirements can be made beforehand of a practice and theory of time-measurement, because it is not methodologically available. Only with the help of clocks which must already always satisfy definite functional criteria can such physical laws be found. On the other hand it is an important result of efforts to establish physics to see that, apart from the homogeneity of their operation, we also expect of our clocks that their measurement results be independent of the velocity of the standard motion (in the sense just explained). If one understands the use of clocks in experimental physics in this way, then naturally, after this, the discovery of the temporal invariance of physical laws given above cannot be considered to be an empirical discovery. It is a simple logical consequence from the conditions which we have incorporated into experimental practice itself, through our use of specific clocks (and in a thoroughly meaningful way, namely with the intention of following specific objectivity criteria). Consequently, in conclusion, we may hold to it as a requirement for the theory and practice of a yet to be established time-measurement that, for time and motion comparison of non-simultaneous events, a standard motion must be found that is homogeneous and with which velocity-invariant temporal propositions can be gained. That these temporal propositions are then also pure, i.e., can only concern the comparison of events with respect to order and duration, would always be implicitly understood. Here, namely, it is merely a question of a logical procedure which goes from propositions about the velocity and form of motions to propositions about the order and duration of their segments by means of abstraction.

2. MOVED BODIES

In accordance with the methodological introductory remarks in Chapter II, we must first begin the terminological fixing of a scientific language with the exemplary introduction of predicates.

Let the predicates 'thing' and 'change' be studied in examples. Examples for things may be: stones, plants, animals, men, the moon. The flight of a bird, the death of a living creature, the coloring of leaves, the flow of a brook, earthquakes and lightning may be called changes. 'Thing' and 'change' are contraries, i.e., the following rules are established:

$$X \text{ is a thing} \Rightarrow X \text{ is no change,}$$

$$X \text{ is a change} \Rightarrow X \text{ is no thing.}$$

In everyday speech it is common, prior to any theory, to speak of moved things. Also in the construction of a theory, 'moved' may be defined by way of example. To be sure, the use as a two-place predicate is immediately intended, which permits a thing 'moved relative to' another thing to be predicated. In addition 'moved relative to' must be a basic predicate, i.e., its introduction is bound to the action schema for moving one thing relative to another. Even here the bounds of a haptic physics are not abandoned. In order not to burden the reconstruction of a rigorous language of mechanics with an exaggerated artificiality, we will already now, in anticipation, replace the predicate 'thing' with the term 'body'. Strictly speaking, this replacement may only be undertaken when, after having introduced protophysical and physical predicates, we can speak of the 'physical context'. In which then as a *façon de parler* 'body' takes the place of 'thing'.

Here when 'body' takes the place of 'thing' as long as 'thing' occurs in a definite context, this procedure is reminiscent of ideation. In distinction to ideation, though, the context here is simply characterized by means of a class of predicates. A 'realization' of bodies would therefore be no meaningful expression. Let us call the procedure demonstrated here by way of example 'supposition'. For any two bodies K_n, K_m let the following hold:

$$\text{(R)} \qquad K_n \text{ is moved relative to } K_m \Rightarrow K_m \text{ is moved relative to } K_n.$$

Thus by regulating the use of the basic predicate 'moved relative to' on the linguistic level, we qualify the designation of reference bodies in the respect that, for any number of moving bodies, any body can respectively act as the reference body.

Following the model of the foundation of geometry, the talk about phenomena of motion shall now be established operatively. If we start with the ability to bring bodies, occurring first as unmanipulated natural objects, into contact and thus to terminologically fix the first possibilities of controlling their surface forms; and then proceed to the ability to systematically change surface forms, thus marking out procedures of manipulation that may be repeated independently from the availability of other forms; and if, finally, we come to speak about the forms so produced as invariant relative to the incomplete carrying out of the procedures of production, we obtain then a system of propositions which is called traditional geometry. Here now we start analogously with the skills of active men – more accurately, of men producing and using instruments – to move bodies, to force a fixed direction upon motions (say by guidance along another body) or also to enforce motion forms. As for the goal of being able to reproduce standard motions for time-measurement, this will depend on marking out repeatable procedures for producing motions artificially. Thereupon we can introduce a way of speaking about the motions produced thusly which is independent, relative to the instructions for repetition, of deficiences in realization. The system of sentences produced in this way represents a theory of motion and shall therefore be called 'kinematics', though, in contradistinction to traditional presentations of a theory of motion, it does not refer to a time parameter and consequently does not refer to clocks.

Firstly we are faced with the task of making distinctions with whose help motions of bodies becomes linguistically and manually or technically controllable; here the first steps in geometry analogous to this can serve as an aid to orientation. (The only difference is, simply, that for a theory of motion we can already have recourse to geometry). The attempt to obtain initial propositions about motions by means of possibilities of control within the scope of comparisons of motion manifests certain difficulties for constructing a kinematics following the model of geometry. In geometry many natural objects are already by themselves sufficiently constant in their form. It is self-evident that one prescribes repeatable control procedures for fixing body forms for, say, stones and not for cumulus clouds. With regards to repeatable measurements, then, the establishment of geometry can also be understood as an effort to refine the pre-scientific-practical criterion of constancy of form into the rigidity of artificially produced measuring instruments in transport. However, for a theory of motion, one cannot similarly fall back upon a rich stock of constant phenomena of motion. While the constancy of form of a large set of natural objects may be naively asserted in a first step, and while it only becomes problematic in highly complex physical or philosophical reflections, the assertion of the constancy of motions is problematic from the outset. What is constant in motions? Since, in speaking about the constancy of motions, we evoke time in the very act of speaking, we must ask, this time, what survives in a motion in such a way that we can succeed in inventing repeatable procedures for comparing motion.

Such reflections may, at first glance, have nothing to do with the foundation problems of natural science or at least be extraordinarily removed from them. However a searching examination of the way in which time-measurement is understood in physics and in the analytic theory of science shows that there are differences of view which go back to the just mentioned foundational reflections. Their clarification therefore is indispensable at this point.

In spite of numerous differences in detail, nearly all theories concerned with time-measurement exhibit a methodical circle when they try to comprehend the constancy of phenomena of motion by means of a description aided by metrical concepts such as velocity or time. Traditionally, when such a sequence is not stood on its head by a more recent relativistic tradition (which explicitly refrains from foundations), kinematics has been constructed after geometry. Actually there is no methodological objection to having recourse to a geometry constructed independently of motion phenomena and theoretical and practical mastery of them. A problem first presents itself with the question, in what respect a geometry must be extended to give rise to a kinematics. Traditionally this has been the time parameter, with whose help, then, velocities, forms of motion, etc. can be defined. However, if one takes into consideration the fact that the time parameter is a magnitude of measure, then one must also speak about motions measuring time. If, finally, this talk of standard motions for time measurement is also itself supposed to fall under postulates of scientific exactness and objectivity, then a theory of motion is already necessary here. In Chapter I various traditional proposals of physics as well as of the theory of science regarding time-measurement were criticized. This critique contained, roughly stated, the assertion that the theories of time discussed do not suffice for a foundation of physics, but rather at the most are a supplementary description of time-measurement disregarding the important aspects of the production and use of instruments. The critique now presented here in connection with the question of the constancy of phenomena of motion in time goes one step further: it asserts that a kinematics cannot be realized by merely adding to geometrical propositions a time parameter or a metric concept 'duration'. Rather kinematics must precede the definition of a metric concept of time.

Obviously this critique is not to be established solely from a consideration of the structure of systems of propositions. Within the domain of verbal definitions alone, the traditional path from geometry to kinematics is of course correct. Rather the critique here aims at the presuppositions which have implicitly entered into traditional views. Among these is to be counted the presupposition that time is a constantly growing parameter, as is expressed most conspicuously in the diagram of the so-called world-lines of the four-dimensional Minkowski-world. According to this diagram every individual motion corresponds to a segment of a world-line. If one would then assert the constancy of motion phenomena in time in the sense of repetitions of equal motions, this would mean considering two segments of world-lines as indistinguishable. Their indistinguishability is to be asserted relative to all available physical parameters, including the time parameter itself, i.e., in particular the lengths and forms of world-line segments must be identical in order to be able to speak of the repetition of an event or a motion. The question of whether a motion or an event is a repetition of an earlier motion or event becomes thereby a question to be decided with the help of measurements. Expressed yet more pointedly: identification of temporal phenomena as recognition of the recurrence of the same phenomenon is possible only with the help of physical theories. Here we have an example (which may be completed by a long series of others) of the fact that physicists (and possibly all natural scientists) examine all subjects falling in the object domain of their science, such as e.g., local motions of bodies, with the means of their theories. For many purposes this is of course a meaningful procedure; not, however, for the purpose of better understanding physics and investigating its presuppostitions.

It is also obvious what implicit presupposition comes to bear here in the procedure of the natural scientist: if there is any constancy of temporal phenomena at all, in the sense of a recognition of the same phenomenon, then this is so by natural law. In other words, because it is implicitly presupposed that nature is controlled by laws (for which due to philosophical objections, many similar sounding alternative formulations have been circulated in the literature), then according to the basic conviction of natural science it is only a matter of recognizing repeated phenomena by means of natural laws. And for this purpose, among others, a time-measurement which would likewise depend on the repetition of equal events or motions such as pendulum or quartz oscillations, a repetition conditioned by natural laws, would serve. If one thereby thinks of Carnap's view discussed in Chapter I that the selection of a standard motion for time-measurement is to be directed toward the greatest possible simplicity of the thus verified physical theories, then a physics working with time-measurement amounts to the search for propositions about (chronometrically) comparable classes of motion phenomena and one is inclined to consider the actions of the physicist measuring time as such a subclass of phenomena of motion. A further consequence of this basic attitude has already been stated: Because the constancy of temporal phenomena such as body motions is considered only in relation to such other phenomena which are likewise, however, conditioned by natural laws, it is not possible 'absolutely', i.e., without reference to 'natural laws', to distinguish forms of motion. What at most appears possible, according to this natural scientific view, is an exemplary definition of forms of motion such as e.g., a definition of periodic motion by pendulum oscillations or of uniform motion by inertial motions of bodies uninfluenced by forces, where the recognition of examples is based on measurements: for pendulum motions, besides measurements, the force field must also remain constant; for inertial motions the sum of the forces affecting the moved bodies must disappear.

Not a few authors propose these interpretations of the historically arisen art of time measuring only because they see no other way. For these authors the foundational problem of time-measurement presents itself specifically in the fact that one can transport standard bodies (measuring rods) but not standard events. To be sure this expresses only a naively realistic misunderstanding: bodies simply occur as natural objects with a sufficient constancy of form; but natural events, once they have occurred, are irretrievably past. The misunderstanding lies in assuming that, *qua* natural objects, bodies are already suitable for comparisons of length, more suitable in any case than are natural events for time-measurement. Nevertheless it is our criteria which make bodies (in favorable cases also bodies as they are by nature) suitable for comparisons of length. Thus the problems of comparisons of length and of duration are strictly analogous: in both cases there must be a transport which requires time, and the problem to be resolved consists in fixing criteria for recurrence of the same phenomenon, i.e., of the same length or the same duration. These criteria must at least *also* relate to artificially produced properties of bodies (e.g., their shape) or of events (e.g., their forms of motion). But the circumstance that we unscientifically identify bodies, as a rule, according to (already present) shape and color, sometimes also according to their length, but identify events mostly in dependence on the thereby moved bodies, is no reason to make a distinction between the comparison of spatially separated lengths and temporally separated events; a distinction of the sort that the former is feasible in good approximation without physics, the latter, however, only with the help of physics.

The methodological discussions of Chapter II should already have shown what alternative view stands behind a constructive foundation of physics, here of time-measurement. It is not natural laws but goal-directed actions that achieve the recognizability of what is equal and which are here therefore the solid foundation for talk about the constancy of temporal phenomena. No physicist can – with whatever instruments – ascertain the indistinguishability of two motions if he has not learned to speak and act in a communicational context. Every action, and *a fortiori* the development of science, already presupposes the ability to recognize situations again and to repeat actions. What can at most then occur within the scope of a natural science is the additional insight that definite products of man are apparently aided by nature, such as e.g., the manufacture of (sufficiently for our purposes) rigid instruments and the reproduction of definite motions such as e.g., partial inertial motions or other standard motions seized upon by clock makers or technicians.

Thus, starting from the elementary abilities which every scientist already carries along with him before he learns his science, the art of time-measurement or its foundation in a theory must here begin with the means of ordinary language. Everyday language, then, supplemented by explicit terminological agreements on account of a possible misunderstanding of it in respect to the words of interest here, must already allow us to describe a repetition of what is the same so exactly that in case of doubt we can conduct a dialogue about the occurrence or non-occurrence of the repetition of an event (possibly with the result that we must make further terminological agreements). The linguistic comprehension of the recurrence of equal events or motions must then be supported by 'objective' control procedures, i.e., by such procedures which may be repeated. These control procedures also presuppose no time-measurement and no physical theorems. Only when finally the goal of physical measurements has been established, must standard events be made transportable (in a formulation analogous to geometry), i.e., criteria for the reproduction of standard events must be formulated. And to avoid methodical circles, these criteria may not be sought with reference to physical theories. This path shall now be further pursued.

Since in the following only *motions* of bodies remain of interest, other possibilities of making statements about bodies will remain out of consideration. Therefore it suffices to represent every body by a mark on its surface.

This mark can be made arbitrarily smaller, thus guaranteeing an indication of place arbitrarily capable of improvement (by indicating the distance of one body from another).

The restriction to *one* mark for every body implies that only 'translational motions', but not however 'rotational motions', will be considered. Also, to terminologically indicate the restriction of statements about bodies to those which hold for a mark on the surface of bodies, in kinematics we will speak of '*point-bodies*'.

Apart from the human ability to enforce on bodies motion relative to other bodies, we can prescientifically make use of the additional ability to guide motions, say in that during its motion a body is kept in contact with another body moved against it. A point-body then describes a 'track' on the surface or an (extended) body. In tracks we distinguish '*trajectories', '*paths' and '*places'. These three predicates are exemplarily introduced for exhibitable bodies and their tracks, thus producing an analogy to the geometric (there to be sure ideatively determined) terms 'line', 'segment' and 'point': The only propositions about *trajectories which are possible are those which concern their shape and position, however not their beginning, end or length; while *paths are supposed to be segments of *trajectories bounded on two sides. Finally *places are boundaries which cut out *paths from *trajectories or which divide *paths into shorter *paths. Thus the demand for the use of '*trajectory', '*path' and '*place' analogous to that of 'line', 'segment' and 'point' yields predicate rules for refining predicates which have been introduced only by way of example. However, these still designate empirically producible objects.

In the procedure of supposition '*trajectory', '*path' and '*place' are replaced by the terms 'trajectoy', 'path' and 'place' (without *), only when geometrical propositions about *trajectories, *paths and *places as well as propositions about the position of point-bodies on these or about the point-bodies' describing a track are under consideration. Consequently by supposition precise terms for physical, more accurately kinematic states of affairs are defined; these terms are ideative in so far as their introduction depends on the ideative definition of geometric terms. Thus propositions about trajectories, paths and places must be understood as ideative norms which become accessible in the procedural instructions of a realization.

The introduction of a further triple of terms, namely of 'motion', 'event' and 'position' proceeds similarly and shall therefore not be presented in extenso. Rules analogous to those for 'trajectory', 'path' and 'place' hold: With motions no propositions about beginning and end are possible; events, however, proceed along definite paths and are thus bounded; as paths are cut out of trajectories by places, so events are parts of motions and are bounded by positions. Consequently the following terms are now available:

point body		
moved relative to		
trajectory	path	place
motion	event	position
(in analogy to		
line	distance	point).

3. COMPARISONS OF MOTION

In the following let the restriction to *linear* motions of point-bodies hold. They can be enforced by guidance along straight trajectories. To carry out comparisons of motion, several terms are introduced by definition:

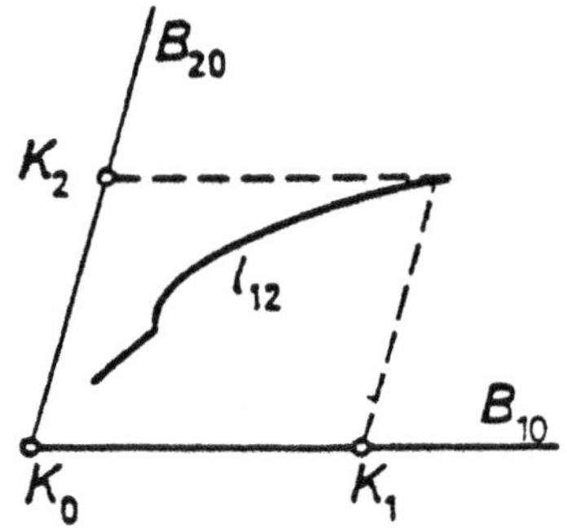

Fig. 1.

(D 1) Two point bodies K_1 and K_2 move relative to K_0 on trajectories B_{10} and B_{20}. The intersection of the parallel to B_{10} through K_2 and of the parallel to B_{20} through K_1 describes a curve relative to K_0; let this curve be called 'guideline l_{12}'. With situations in which the motion relative to the same reference body has not been stipulated for all point-bodies, let us write $l_{12;0}$ to avoid misunderstandings. The index located after the semicolon indicates the reference body.

As a further restriction let, for the following discussion, the requirement always be fulfilled that the motions considered be 'simple', i.e., they contain no reversal positions. Every place of a trajectory will be run through only once.

After 'moved relative to' was introduced exemplarily as a basic predicate, thus assuring at the same time that the command to move a body relative to another represents an intelligible procedural instruction, the state of rest is marked out by definition.

(D 2) A point-body is *'at rest'* (synonoymously: 'resting', 'it rests') relative to K_0 and relative to an event v_{20}. K_1 assumes a *'continuous position of rest'* at place s, which lies at the intersection of trajectory B_{10} with the extension of the parallel segment of the guideline. In abbreviated form this is written: $K_1 \, r \, K_0, v_{20}$.

The predicate 'at rest' is thus used as a three-place predicate, while 'moved relative to' is used only as a two-place predicate. This asymmetry, which implies that to assert the predicate 'resting' and to deny the predicate 'moved' of a body is not equivalent, appears first to stand in

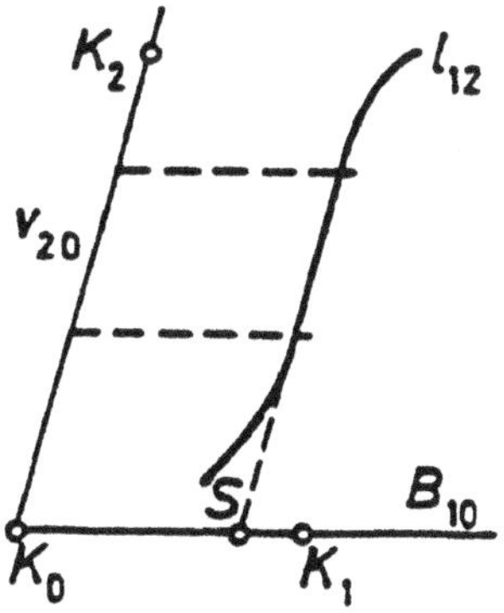

Fig. 2

conspicuous discrepancy to ordinary usage, where 'resting' and 'moved' are contraries. The context of ordinary language talk about rest and motion assures the intelligibility of the propositions in question, since in this context reference bodies and comparison motions are implicitly contained. To be sure, where intelligibility is not attained by means of implicit reference to comparison motions and, in particular, where one agrees to an argument for or against the legitimacy of designating a body as resting or as moved, ordinary language also exhibits the stated asymmetry:

Deciding whether, in an individual situation, a real body is at rest or is moved requires an observation of the body over a certain duration. (This statement is intelligible with a pretheorietic use of 'duration'.)

If now the elapse of the duration may not be defended by alluding to the human time-sense, rather may only be defended by referring to what is 'externally observable', then the distinction between rest and motion clearly exhibits itself: With the moved body a reference to its own change of place suffices to demonstrate the elapse of a duration, while with the resting body a comparison event must be named which yields an 'objective' criterion for the fact that time has passed.

Within the framework of the foundation of a chronometry this distinction may be comprehended yet more simply: One can speak of motion only if at least two point-bodies are considered. On the other hand for propositions about the rest of a point-body at least three point-bodies are necessary. The 'instantaneous position of rest' presents a limiting case of the continuous position of rest:

(D 3) Two point-bodies K_1 and K_2 move relative to K_0 on trajectories B_{10} and B_{20}. If there is a tangent t to the guideline l_{12} which is parallel e.g., to trajectory B_{20}, then K_1 would have an *'instantaneous position of rest'* at place S, the intersection of t and B_{10}. This is abbreviated as: $K_1\ r\ K_0$, s_{20}, where s_{20} is the position assumed by K_2 at the place s_2.

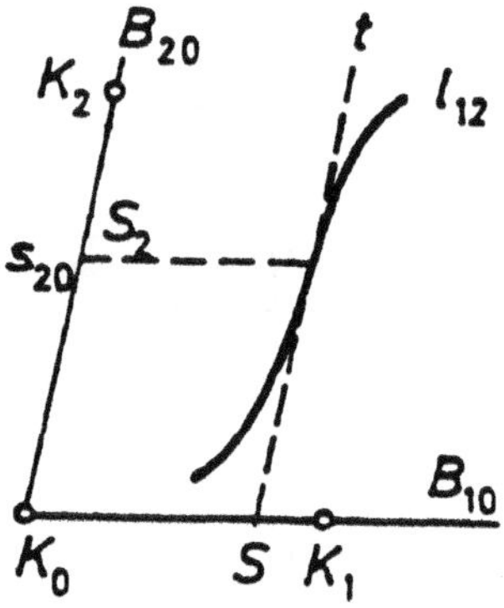

Fig. 3

From (D 2) and (D 3) it follows that the rest of one point-body can indeed be inferred from the motion of another, namely with a suitable course of the guideline; however the inference in the converse direction is not allowed. In (D 2) the point-body K_2 can itself assume positions of rest inside of v_{20}; these then are ascertainable only relative to the motion of a further point-body K_3. Thus if in (D 2) and (D 3) it were to be asserted of the point-body K_2 that it is moved, then this should not terminologically exclude the possibility of there also being positions of rest in this motion.

The straight guideline in particular provides a productive criterion for a purely geometric comparison of motion.

(D 4) K_1 and K_2 move relative to K_0 on non-parallel trajectories. Their motions b_{10} and b_{20} are called *'similar'* if the guideline l_{12} is straight (in symbols: $b_{10} * b_{20}$).

The following definitions serve to simplify the further presentation:

(D 5) Point-bodies K_1 and K_2 move on trajectories B_{10} and B_{20} relative to the point-body K_0 and produce the guideline l_{12}.

Two events v_{10} and v_{20} are called *concomitant events* if they belong to the same segment of the guideline l_{12}.

Two positions s_{10} and s_{20} are called *concomitant positions* if they belong to the same point of the guideline l_{12}.

Paths of concomitant events are called *concomitant paths*.

Places of concomitant positions are called *concomitant places*.

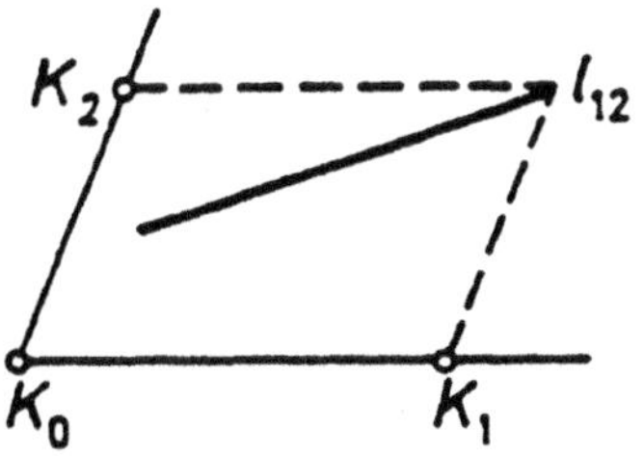

Fig. 4

Therewith (D 4) may be extended to events:

(D 6) Two concomitant events are called *similar* if the segment of the guideline belonging to them is straight.

Finally, relative acceleration or retardation of a motion viv-à-vis another is still ascertainable.

(D 7) Two point-bodies K_1 and K_2 move relative to K_0 on the trajectories B_{10} and B_{20}. Let motions b_{10} and b_{20} be divided into arbitrary events v_{10}^n and their concomitant events v_{20}^n. The indices n may grow in the direction of motion. Thus a division of the trajectories B_{10} and B_{20} into parts of concomitant paths w_{10}^n and w_{20}^n is given.

The motion b_{10} *outruns* the motion b_{20} when for every such division the following holds:

$$|w_{10}^n| : |w_{10}^{n-1}| > |w_{20}^n| : |w_{20}^{n-1}|.$$

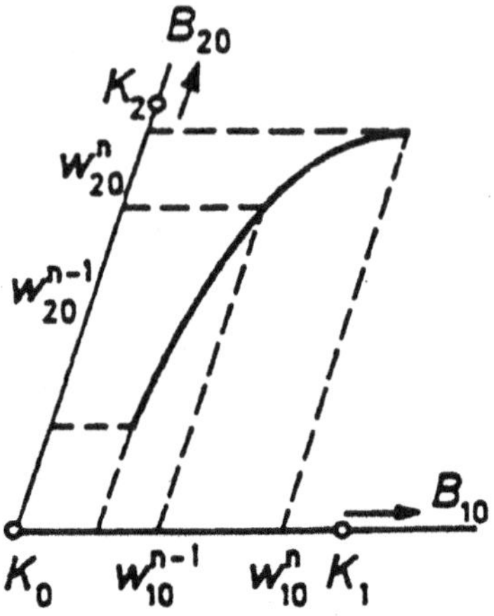

Fig. 5

Let the converse relation to outrunning be called *lagging behind*.

We have thus provided a terminological apparatus that permits the construction of a clock-free kinematics. To be sure, this kinematics can contain only theorems about comparisons of motions which can be carried out on 'simultaneous' (in the prescientific sense) motions. Indeed this does not allow us to conclude that such a theory could not already provide a sufficient basis upon which empirical research may be set in motion. Before giving some examples of this, we shall derive several theorems from the definitions (D 1) to (D 7). As regards the selection of these theorems we should say that no sort of completeness of a system of theorems is striven for here, which system would then be a closed clock-free kinematics. The clock-free comparison of motion has no sort of practical significance of the sort that we could now recommend to, say, an experimental physicist that he should henceforth refrain from using clocks in some experiments. The significance in principle, however, which lies in the proof that, even without clocks, one can obtain qualitative as well as quantative empirical propositions about motions, may be sufficiently illustrated by giving some examples. Thus the derivation of theorems from definitions will be restricted to those which will be necessary in the later construction of chronometry as well as to others which are indeed used as elementary kinematic theorems in current textbooks of mechanics, but which have never been proven.

3.1. *Theorems on Similarity.* [18]

Similarity of motions was defined by the condition that the guideline be straight. Equivalent to this condition is the following proposition: Pairs of concomitant paths of similar motions have a constant ratio of length. In this form, which pre-supposes that a measurement of length is available, the definition of similarity offers the advantage of enabling us to easily reduce propositions about motions or events to propositions about rational numbers; in the following this will allow us to shorten proofs by only carrying out this reduction.

Symmetry of similarity holds trivially, as one sees immediately. If a set of quotients of path lengths is one and the same rational number, then the reciprocals of these quotients are also always one and the same rational number. Thus the following holds:

$$b_{no} * b_{mo} \rightarrow b_{mo} * b_{no}.$$

Transitivity is proved just as easily: Let three point-bodies K_1, K_2 and K_3 move relative to K_0 on the non-parallel trajectories B_{10}, B_{20} and B_{30}. Let $b_{10} * b_{20}$ and $b_{20} * b_{30}$ hold. Thus for all divisions of the three trajectories into concomitant paths w_{10}^n, w_{20}^n and w_{30}^n, the constancy of the ratios of the path-lengths $|w_{10}^n| : |w_{20}^n|$ and $|w_{20}^n| : |w_{30}^n|$ is presupposed; from this the constancy of the path-length ratios $|w_{10}^n| : |w_{30}^n|$ follows directly. With the help of symmetry and transitivity, reflexivity may be proven for all motions for which there are similar motions. This logical restriction is trivial here, since it may be considered as always fulfilled for forced motions.

Thus we have shown the reflexivity, symmetry and transitivity of similarity; 'similar' is an equivalence relation. Theorem (S 1) holds accordingly:

(S 1) a) $b_{nm} * b_{nm}$,

b) $b_{nm} * b_{om} \rightarrow b_{om} * b_{nm}$,

c) $b_{nm} * b_{om} \wedge b_{om} * b_{pm} \rightarrow b_{nm} * b_{pm}$.

As rule (R), the symmetry of the relation 'moved relative to' was required. This rule implies the identity of events v_{nm}^k and v_{mn}^k. From this follows the equality of the path-lengths $|w_{nm}^k|$ and $|w_{mn}^k|$. which can also be read as the constancy of the path-length ratio $|w_{nm}^k| : |w_{mn}^k|$; i.e., the theorem (S 2) holds:

(S 2) $b_{nm} * b_{mn}$.

Theorem (S 2) describes the same state of affairs as (S 1a); this also explains what the similarity of a motion to itself means materially: the criterion of similarity does not distinguish reference bodies. The next step consists in an application of rule (R) to comparisons of motions, thus to situations in which three or more point-bodies participate. It will allow the introduction of a way of speaking about the superimposition of motions. In particular from this we may derive a theorem of classical mechanics, a theorem mostly cited there under the title 'parallelogram of velocities'.[19]

Two point-bodies K_1 and K_2 move relative to K_0 on the non-parallel trajectories B_{10} and B_{20}. Let their motions be similar, i.e., $b_{10} * b_{20}$. Let the guide-line l_{12} pass through K_0.

Now to the point-bodies K_0 and K_2 rule (R) is applied, which implies the identity of b_{20} and b_{02}. If now the motion b_{10} of point-body K_1 is no longer considered relative to K_0 but also relative to K_2, we must then explain what propositions hold for the 'new' motion b_{12}.

Let motions b_{10} and b_{20} be divided into events v_{10}^1, v_{10}^2, ... and their concomitant events v_{20}^1, v_{20}^2, ... Then, according to rule (R), the identity of v_{20}^n and v_{02}^n holds. If one observes

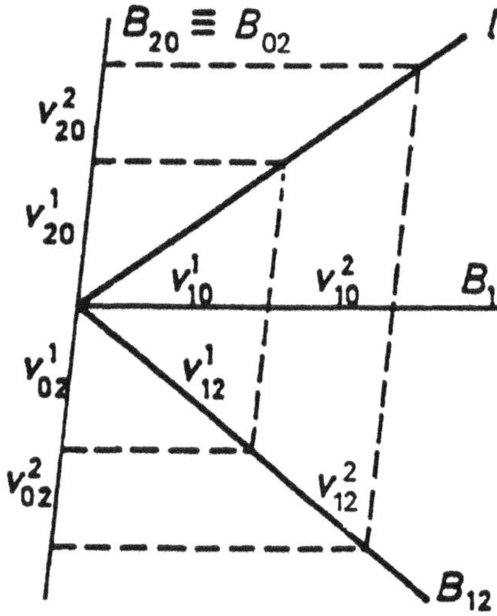

Fig. 6

that rule (R) requires the equivalence of two representations of one and the same state of affairs, then it is also presupposed here that the state of affairs supposed to be presented with new propositions is the same as in the original representation. That means, in particular, that the triangles K_0, K_1, K_2 are congruent in 'simultaneous descriptions' in both representations. (To be sure the word 'simultaneous', used here for clarity, may be avoided as well and, to that extent, does not violate methodological order. It suffices here to assert the congruence of the triangles in the respective end positions of the events v_{20}^n and v_{02}^n.)

Then an elementary geometrical reflection shows that the paths of the events v_{12}^n are diagonals of parallelograms, the sides of which are deternimed by the paths of the events v_{02}^n and v_{10}^n.

This result is to be interpreted as follows: A point-body K_1 moves relative to K_0 and K_0 itself moves relative to K_2. The trajectories of the motions b_{10} and b_{02} are given. Then the trajectory of the motion b_{12} can be constructed according to the parallelogram rule just obtained. Thus a 'resulting track' has been found which arises for K_1 by 'superimposition' of motions b_{10} and b_{02}. A corresponding proposition holds for concomitant paths. Propositions about the motion resulting from the superimposition of two motions may thus be obtained a priori and, in addition, without clocks by reduction to a simultaneous, geometrical comparison of motion. As trivial as the present derivation may appear, it may nevertheless be taken as a proof of the theorem presupposed but strictly unfounded in classical physics, namely that velocities add up vectorially. (To solve this problem it was not once necessary to introduce a term 'velocity'!). Moreover it is easy to see that the reduction of the superimposition of two motions to a clock-free comparison of motions also furnishes, with a suitable introduction of the term 'force' (which, to be sure, will not be possible without clocks), the vectorial addition of forces.

The result of the reflection of how a superimposition of motions can be understood must again be considered in connection with the normative character of the concepts employed. 'Straight trajectory' was introduced as an ideative concept to which a realization procedure corresponds. Thus to speak of trajectories implies that it must be possible to produce a rail (Schiene), a groove, etc., to guide a body. Propositions about ideal trajectories must then be satisfiable for their realizations. How is the construction of a resulting trajectory to be incorporated into this schema?

Here first the above assertion that rule (R) serves only the different representability of one and the same state of affairs must be made precise. While for two bodies this rule in fact allows the same state of affairs to be represented in two ways (either of the two bodies can be considered as the reference body), in the case of three moving bodies we can speak of identical states of affairs only in a restricted sense, in spite of an exchange of reference bodies: Independently of the choice of the reference body and consequently independently of the representation, the situation of the bodies to one another (in simultaneous positions) is the same. However, as for the trajectories, different realizations correspond to equivalent representations (equivalent namely insofar as they indicate the situation of bodies in simultaneous positions to be the same). On one hand the pair of trajectories B_{10} and B_{20} must be realized, on the other, the pair of trajectories B_{02} (identical with B_{20}) and B_{12}. The availability of different but equivalent representations presupposes – we may formulate it as a result – a preliminary decision as to which state of affairs must appear the same in differing representations, where here 'representation' is not to be understood merely verbally or formally but also involves the realization. Thereby we have not decided whether in dynamics, where 'constraining forces' (*Zwangskräfte*) must also be allowed for in connection with trajectories, the same invariance can arise in respect to an exchange of reference bodies.

Up to now the superimposition of two motions has been considered under the supplementary condition that the guide-line l_{12} of similar motions b_{10} and b_{20} passes through the reference body K_0. The extension to cases in which this condition has not been fulfilled shall be demonstrated only indirectly here. In Figure 7 which shows both representations (K_0 is the reference body, K_2 is the reference body), the triangles K_0, K_1, K_2 and (K_0), (K_1), (K_2) (for the earlier positions), have been drawn in; both triangles have been presupposed to be congruent to each other.

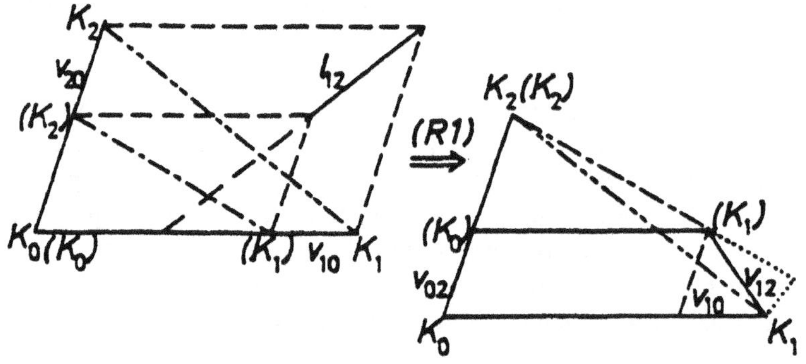

Fig. 7

The free choice of a reference body from three point-bodies K_m, K_n, K_o moving relative to one another, a choice made according to rule (R), leads to the 'addition theorem for simultaneous motions':

(S 3) $\qquad b_{mn} + b_{no} = b_{mo}$.

The abbreviated formulation, which can be read as 'the superimposition of the motions b_{mn} and b_{no} results in the motion b_{mo}, signifies nothing more than a procedural instruction for assertaining for concomitant events with the help of a parallel construction, the path that a point-body describes after both other point-bodies have been exchanged in respect to the choice of them as reference bodies.

Finally, the addition theorem may also be extended to non-similar but simultaneous motions: For this motions are divided into differential partial events, where for two respectively concomitant events with paths whose lengths converge toward zero we can assume similarity. The addition theorem holds accordingly for all differential partial events of a motion and consequently furnishes an instruction for constructing a resulting motion as a whole.

Since, in this work, a clock-free kinematics is carried out only to a stage necessary for establishing time-measurement, the addition theorem for non-similar motions will not make its appearance and thus also need not be formulated exactly. As one immediately sees, the superimposition of two non-similar motions, but motions on straight trajectories, leads to a resulting trajectory that is no longer straight. However a clock-free comparison of motions shall not be extended here to non-linear trajectories, since it furnishes no additional contribution to understanding the foundations of chronometry.

However, let us here refer to a peculiarity of the situation represented in Figure 7: For K_1, under the supplementary condition that the guideline l_{12} does not pass through the reference body K_0, theorem (S 3) leads to a trajectory that is indeed straight, but which does not pass through the new reference body K_2 (see right side of Figure 7). If on the other hand, one considers the motion of the point-body K, on the trajectory which in all positions coincides with the straight line through K_1 and K_2, then Figure 7 shows that the angle between this trajectory and B_{20} does not remain constant during the motion. Further pursuit of this question would result in an addition theorem analogous to (S 3) which determines the superimposition of a rotary motion with a linear motion in a radial direction. In this connection we could carry out a transition to polar coordinates without using clocks; however, this shall not be done here.

After the superimposition of two motions has become resolvable into a simple comparison of two motions referred to the same body, we must ask whether the resulting motion is still similar to the motion of the other body. This question may be generally understood as: Is the similarity of motions invariant relative to rule (R)? Only a proof of this invariance secures the fact that propositions about similar motions are independent of the choice of the reference system (in clock-free kinematics, for the present, only of the choice of the reference body); in other words, it assures that we need fear no influence of an accidentally favorable choice of the reference body on the validity of general propositions about similar motions.

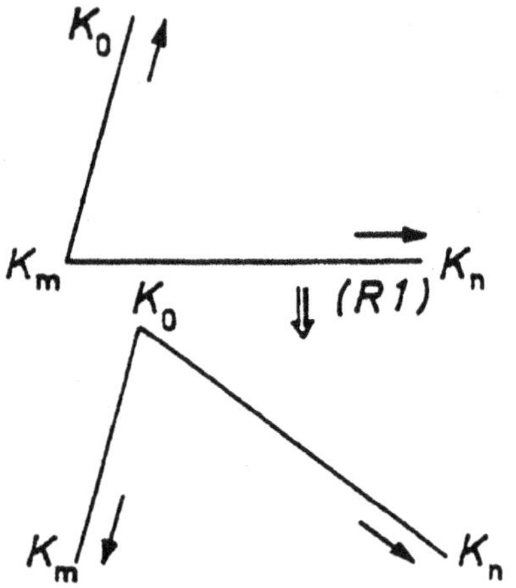

Fig. 8

The invariance of similarity relative to rule (R) may be formally comprehended as theorem

(S 4) $$b_{nm} * b_{om} \wedge b_{nm} + b_{mo} = b_{no} \rightarrow b_{mo} * b_{no},$$

which can be proved by elementary geometric means (for illustration of the indices of (S 4) see Figure 8).

In the starting situation (see Figure 9) with the conditions $b_{10} * b_{20}$, guideline $l_{12;0}$ passes through reference body K_0; rule (R) is applied to K_0 and K_2. For K_1, in accordance with (S 3), the trajectory B_{12} results. Then, in accordance with (D 1), the definition of guideline, guideline $l_{01;2}$ is straight.

Thus (S 4) is reduced to the following geometrical problem (see Figure 10): Do the corners A, F and F' of the parallelogram $A\,E\,F\,G$ and $A\,E'\,F'\,G'$ lie on a straight line when the following assumptions hold (abbreviation: $\underline{AD}$ means 'segment $\underline{AD}$'):

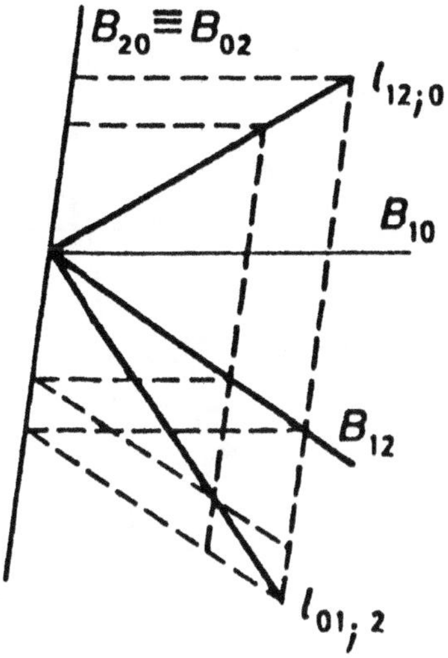

Fig. 9

$\underline{AD} = \underline{AE}$;	$\underline{AD'} = \underline{AE'}$	(from (R))
$\underline{AD} = \underline{BC}$;	$\underline{AD'} = \underline{B'C'}$	(from (D 1))
$\underline{AE} = \underline{BG}$;	$\underline{AE'} = \underline{B'G'}$	(from (S 1))
$\underline{AE} = \underline{FG}$;	$\underline{AE'} = \underline{F'G'}$	(from (D 1)).

Points A, D, D', E and E' lie on a straight line. (This assumption is fulfilled due to the restriction to straight trajectories and the supplementary condition that $l_{12;0}$ passes through K_0.)

Then the following holds: Points C, B, G, F lie on a straight line. Points C', B', G', F' lie on a straight line.

However, therewith points A, F, and F' also lie on a straight line; which was what was to be shown. Thus $b_{12} * b_{02}$ holds.

By means of a simple auxiliary construction we may also reduce the case where the supplementary condition ($l_{12;0}$ passes through K_0) is not fulfilled to the indicated situation, so that the invariance of similarity relative to (R) is independent of this condition.

In (S 1)c) transitivity of similarity for three point-bodies, all moving relative to the same point-body, was proved. In a later proof, transitivity must also be presupposed for the case where, in a system of several bodies, different reference bodies occur simultaneously. This extension to comparisons

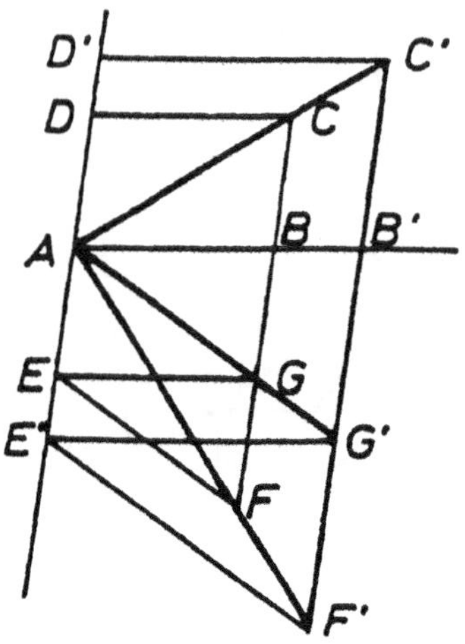

Fig. 10

of motion which are arbitrary in regards to the choice of reference body corresponds moreover to the situation in which the physicist experimenting with a clock finds himself. There the motion of the clock hand is not referred to the same body relative to which the motion being investigated with the help of the clock is determined.

First let the following arrangement be given: Three point-bodies K_1, K_2 and K_3 move relative to K_0 and the following holds: $B_{10} \equiv B_{20}$ (i.e., K_1 and K_2 run on the same trajectory), $B_{10} \not\equiv B_{30}$.

Moreover let the following hold: $b_{10} * b_{30}$ and $b_{20} * b_{30}$ (see Figure 11). Then, the assertion to be proven is that the motion of K_1, relative to K_2 is similar to the motion of K_3 relative to K_0, i.e., $b_{12} * b_{30}$.

The proof again makes use of the reduction of propositions about comparisons of motions to propositions about rational numbers.

Let trajectories B_{10} or B_{20} and B_{30} be divided into concomitant paths w_{10}^n, w_{20}^n and w_{30}^n. Then, by the assumed similarity, the following holds: The path-length ratios $|w_{10}^n| : |w_{30}^n|$ and $|w_{20}^n| : |w_{30}^n|$ are constant for all n.

The differences of the path-lengths w_{10}^n and w_{20}^n are equal to the path-lengths w_{12}^n for all n, where w_{12}^n are paths concomitant to the paths of the assumed division. Then from

$$|w_{10}^n| = a\,|w_{30}^n| : |w_{20}^n| = b\,|w_{30}^n| \text{ } (a, b \text{ rational numbers})$$

$$|w_{12}^n| = |w_{30}^n|\,\|a - b\| \text{ } (\|x\| \text{ means: absolute value of } x)$$

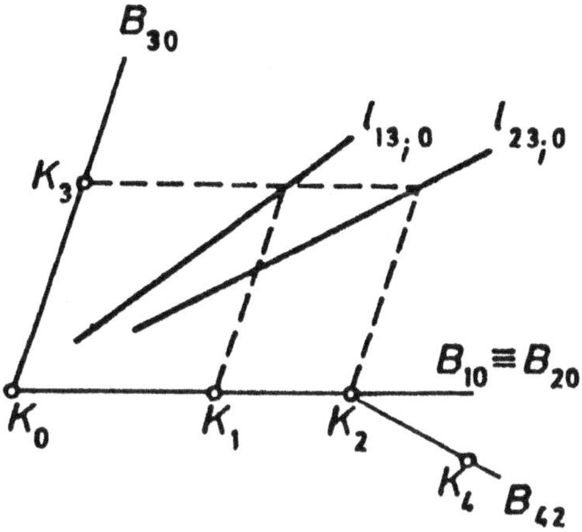

Fig. 11

follows and thus for a comparison of the motions b_{12} and b_{30}:

$$|w_{12}^n| : |w_{30}^n| = \|a - b\| \text{ for all } n, \text{ i.e., } b_{12} * b_{30}.$$

The addition theorem (S 3) also defines 'addition' and 'subtraction' for those paths which do not lie on one trajectory. With this extension, the theorem just proved may be written as

(S 5) $\quad b_{nm} * b_{om} \wedge b_{pm} * b_{om} \to b_{np} * b_{om}.$

The relation $b_{np} * b_{om}$ obtained in (S 5) is the result of a formal extension of the possibilities of comparing motions. Up to now similarity was only defined for motions occurring with regard to the same reference body. Therefore we should ask whether the formal extension to arbitrary comparisons of motion, i.e., with a total of at least four point-bodies, can also be an ideative norm of the type that one could state how a realizatum, i.e., some instrument or other which permits such a motion comparison would look. Only when this question is answered positively, is theorem (S 5) justified within the scope of a protophysical theory.

The admissibility of extending 'similar' to a predicate for (in respect to their reference body) arbitrary motions, can be shown by giving an auxiliary construction which reduces the extended comparison of motion to those arrangements of bodies which underlie the definition of similarity.

Starting from the same situation as in (S 5) (see Figure 11) a fifth point-body K_4 can be added which moves relative to K_2 on a trajectory B_{42}. B_{42} forms with B_{20} an arbitrary, but during the motion, constant angle. According to assumption, $b_{30} * b_{20}$ holds. If now for the motion of K_4 the conditions $b_{42} * b_{20}$ and $b_{42} * b_{21}$ are fulfilled, then, by (S 2) and (S 1)c), the following holds:

$$b_{30} * b_{20} \wedge b_{20} * b_{42} \wedge b_{42} * b_{12} \rightarrow b_{30} * b_{12}.$$

The premises contain only motion comparisons of the original type so that, in regard to realizability, (S 5) contains no other presuppositions than does the definition of similarity.

Finally yet one further theorem about similar motions will be needed:

(S 6) Let the motions of the point-bodies K_1 and K_2 relative to K_0 be similar, i.e., $b_{10} * b_{20}$. If point-body K_1 at place S_1 has an instantaneous (continuous) position of rest, then K_2 at S_2 (S_2 is the concomitant position of S_1) also has an instantaneous (continuous) rest position.

The proof will first be carried out for the continuous rest position. Thus let K_1 have in S_1 a continuous rest position, i.e., there is an event v_{30} of the motion b_{30} of a further point-body K_3, so that the guideline l_{31} is parallel to trajectory B_{30} in the segment which belongs to v_{30}.

Assume that K_2 relative to v_{30} is not at rest, i.e., that the guideline l_{23} is not parallel to trajectory B_{30} in the segment belonging to v_{30}.

Then there is an event v_{20} of the motion b_{20}, relative to which K_1 in S_1 assumes a continuous position of rest, i.e., guideline l_{12} is parallel to trajectory B_{20} in the segment belonging to v_{20}.

Then from the similarity of motions b_{10} and b_{20} it holds that the total guide-line l_{12} is parallel to trajectory B_{20}, i.e., that the body K_1 is not moved at all.

This special case, that even the total rest of a point-body relative to the motion of another point-body can be considered as 'similar', shall be excluded by a supplementary agreement. For ascertaining that the condition of rest is similar to any arbitrary motion, demonstrates the uselessness of this limiting case for comparisons of motion. From now on then let the definition of similarity apply only to the case in which the guide-line is parallel to neither of the two trajectories.

From this additional condition, which implies, for the still open proof, that K_1 rests relative to no event v_{20}^x of the motion b_{20}, and from the assumption that K_2 does not rest relative to v_{30}, a contradiction to the assumption

$b_{10} * b_{20}$ follows. Thus K_2 rests relative to v_{30}, whereby (S 4) is proven for the case of the continuous rest-position of K_1.

In definition (D 3) of the instantaneous rest-position, the existence of a tangent to the guide-line parallel to one of the trajectories was required. This definition is also applicable to the case in which events are compared with one another and in which the tangent touches at the endpoint of the segment of the guide-line belonging to the events. For the sake of brevity (S 6) will be proved for the corollary condition where K_1 assumes an instantaneous rest-position at the beginning of an event.

As one immediately sees, this also comprises the general case in which the instantaneous position of rest is assumed within an event or motion.

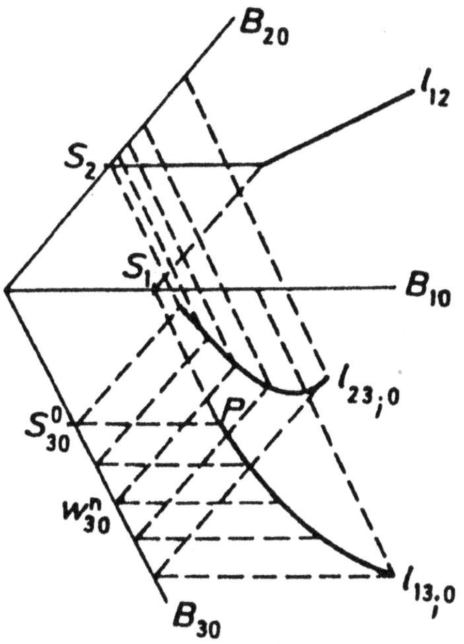

Fig. 12

Let the trajectory B_{30}, starting from the place S^0_{30}, be divided into paths w^n_{30} of equal length. Via the guideline l_{13} the concomitant paths w^n_{10} may be determined on trajectory B_{10}. Then, from the condition that at P (see Figure 12) l_{13} has a tangent parallel to B_{30}, it follows for the paths w^n_{10}.

From a sufficiently fine division of B_{30}, for an $n > N$, where N is to be suitably chosen in dependence on the course of the guideline l_{13} and the fineness of the divisions, the path-lengths $|w_{10}^n|$ with decreasing n form a sequence converging towards zero (*Nullfolge*).

By the similarity of the motions b_{10} and b_{20}, the sequence of path-lengths $|w_{20}^n|$ (w_{20}^n are the concomitant paths to w_{10}^n) with decreasing N is also a sequence converging towards zero.

After the concomitant places S_{20}^n on B_{20} to the places S_{30}^n on B_{30} are thus ascertained via guide-lines l_{12} and l_{13}, the guideline l_{23} may be constructed; l_{23} allows a comparison between b_{20} and b_{30}. Obviously l_{23} must also have a tangent parallel to B_{30} at the point which belongs to S_{30}^0. But consequently it is proven that K_2 assumes an instantaneous position of rest at S_{20}^0, which is what was to be shown.

By the already given extension of motion comparisons to motions with arbitrary reference bodies, (S 6) also holds under the extended condition that the motions of K_1 and K_2 are not referred to the same point-body.

The terminological distinctions made up to this point and the theorems logically following from them represent an apparatus suitable for already tackling in elementary dynamics a problem which is hardly discussed within the scope of empirically understood classical physics, even though one frequently makes use of a preliminary resolution of this problem: the continuity of changes in velocity. To be sure this is not unconditionally a fault of the physicist, since considerations of continuity, which in connection with protophysical basic magnitudes amount to a discussion of the conceptual foundations of measurement, could not otherwise be furthered by empirical contributions.

3.2. *Continuity of Changes of Velocity.*

The question whether *a priori* and possibly even clock-free propositions about velocity-changes are formulatable and finally arguable requires a preparatory clarification. First it may be surprising to wish to formulate propositions about velocity changes in a clock-free kinematics, in which a term 'velocity' has not yet occurred. If this is at all possible, then such propositions will be formulatable only with the help of additional distinctions which, to be sure, are encountered in simultaneous motion comparisons. According to the preparation accomplished up to this point, one can see that these distinctions can be expressed only in two-place predicates for the example of *momentary* comparisons of velocity. However, this also already sets the limits within which the following discussion will move.

On the level of pre- or extra-scientific speech, agreement is easily reached in that velocity of bodies can only change 'continuously'. By this is meant that (1.) even a change in velocity demands 'time', and that (2.) on the assumption that one has already assigned numbers to velocities, a body which assumes two different velocity values v_1 and v_2 must also assume every value between v_1 and v_2. (Mathematically, to be sure, continuity is not equivalent to the validity of the law of intermediate values (interpolation).)

The plausibility of this proposition stems surely from the everyday experience, fully accessible to everyone, that bodies are inert, or what is synonymous, that bodies do not let themselves be accelerated instantaneously. Even where one leaves the level of prescientific distinctions, one could expect, above all, a dynamical foundation for the continuity of velocity changes. In assuming that all bodies have an inert mass different from zero and that only finite forces are available for accelerating bodies, it follows from the Newtonian definition of 'force' that, to change the velocity of a body, a duration different from zero is required.

Without having to lose ourselves in details, we may nevertheless indicate why a foundation carried out in such a way cannot be sound. The definition of 'force' as the product of inert mass and acceleration is only intelligible if a continuum of velocities has already been presupposed as defined. Otherwise 'acceleration' cannot be introduced as change of velocity. But this task falls to kinematics.

Thus it appears unavoidable in this context to go into the continuum problem, a problem as old as human efforts towards exact propositions about nature. To be sure, this problem received an exhaustive treatment already with Aristotle; nevertheless one cannot speak of a reverberation of his reflections in classical or in modern physics.

Here, however, we shall not tackle the continuum problem. By setting out in the reflections made here from a position in which Euclidean geometry is presupposed as a well-founded and finished theory, we also assume that a satisfactory explanation of the continuity of lengths, understood in Aristotelian terms as divisibility into an again divisible length, has already been accomplished. Where geometry is understood as part of a protophysics, it is a geometrical and not a kinematical question whether a body moving along a trajectory from one place to another also runs through all the intermediate places of this trajectory.

The introduction of the terms 'place' and 'position' and in particular the predicate rules supposed to hold for these terms show, at any rate, that the continuous structure of trajectories and motions was presupposed there. For this reason the deferred question of how velocity changes can be comprehended linguistically can be answered without further reflection on the problem of continuity, where to be sure a part of the presuppositions is delegated to geometry and therefore shall not be handled here.

After these preliminary remarks, the available terminology must first be sharpened. The term 'instantaneous position of rest' was defined by the condition that the guide-line has a tangent at the corresponding point that is parallel to one of the trajectories. To speak of 'instantaneous rest position' shall now also be possible for the following special case, for which no conventions have yet been established heretofore in the definitions.

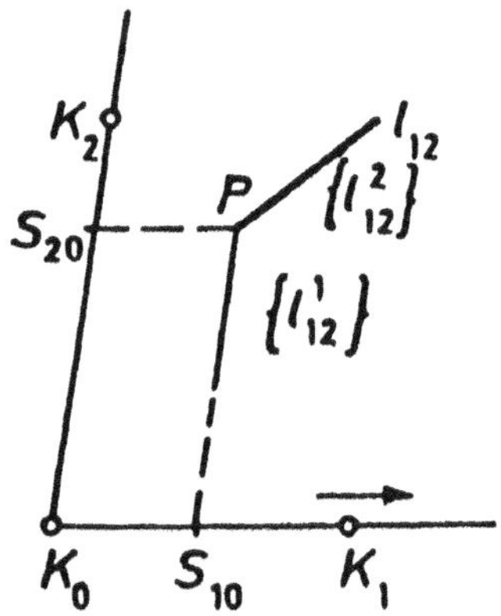

Fig. 13

(D 8) Let guideling l_{12} of the point-bodies K_1 and K_2 moving relative to K_0 have the course shown in Figure 13, i.e., let K_1 have a continuous rest position is S_{10} and outside of S_{10} let it be moved similarly to the motion of K_2. Then, for the place S_{20}, which belongs to the bend point P of the guide-line and at which K_2 assumes the position s_{20}, the following holds (in addition to $K_1\, r\, K_0, v_{20}$) : $K_1\, r\, K_0, s_{20}$, i.e., K_1 has relative to s_{20} an *instantaneous rest position*.

In other words, point P of the guideline belongs per definitionem to the guideline segment $\{l_{12}^1\}$, the segment $\{l_{12}^2\}$ is *open* in the direction towards P. *Continuous* rest is thus always established relative to *closed* comparison events. This stipulation is justified by its compatibility with the hitherto existing terminological specifications. In contrast to the alternative possibility of assuming the comparison event of continuous rest to be open, the solution chosen here is compatible with (D 3), the definition in 'instantaneous rest position'. Rest positions are defined relative to closed 'occurrences', where it shall not depend on 'how long' a body is at rest. This relationship of instantaneous and continuous rest has a deeper reason than say only a formal-aesthetic one. Trivially instantaneous rest may not be defined other than as 'closed'. If one then wishes in contrast to this, to define continuous rest as open then one would fall into a conspicuous contradiction not only to actual usage recognized by everyone, but also to the usage to be normatively reconstructed; that is if the definition itself does not become entangled in contradictions or does not produce the most abstruse special cases: If the comparison event v_{20} of the continuous rest of K_1 were open, i.e., were K_1 to assume, relative to the end positions $s_{20}^{1,2}$ of the comparison event v_{20}, not at (*per definitionem*: instantaneous) rest positions, then one could only call the 'state of motion' of K_1 relative to $s_{20}^{1,2}$ undefined. The goal of enabling rigorous comparisons of motion requires the abolition of such undefined 'states of motion'. Then no other choice remains (if one would not invent a 'third state of motion') but to designate K_1 as moved relative to $s_{20}^{1,2}$. This solution would imply saying that a body, for which it has been stipulated that it is in rest over a duration, is suddenly moved and indeed 'at a time' in which it still assumes the same place. (The expression 'at a time' ought not to be irritating here. It is eliminable by propositions about K_2.) Consequently we may assert that in regard to the consistency of kinematic terminology it is justifiable to define continuous rest relative to closed comparison events; this means nothing other than assuming instantaneous rest positions 'temporally at the beginning and end' (thus relative to the motion of K_2) of a continuous rest position.

After this extension of terminology the following theorem may be proved:

(S 7) Let the guideling l_{12} of the point-bodies K_1 and K_2 moved relative to K_0 have the course shown in Figure 14, i.e., let the following hold: $v_{10}^1 * v_{20}^1$, $v_{10}^2 * v_{20}^2$. S_{10} and S_{20} are concomitant places. The point P belonging to them lies at a bend (*Knickstelle*) of the guideline.

Then, so states the assertion to be proved, K_1 assumes a rest position in S_{10} and K_2 assumes a rest position in S_{20}, i.e., a motion b_{70} of a point-body K_7 can be produced so that the following holds:

$$K_1\, r\, K_0, s_{70} \quad \text{and} \quad K_2\, r\, K_0, s_{70}$$

or

$$K_1\, r\, K_0, v_{70} \quad \text{and} \quad K_2\, r\, K_0, v_{70}.$$

Proof: First new point-bodies satisfying the given conditions are introduced: this leads to an auxiliary construction as is represented in Figure 15.

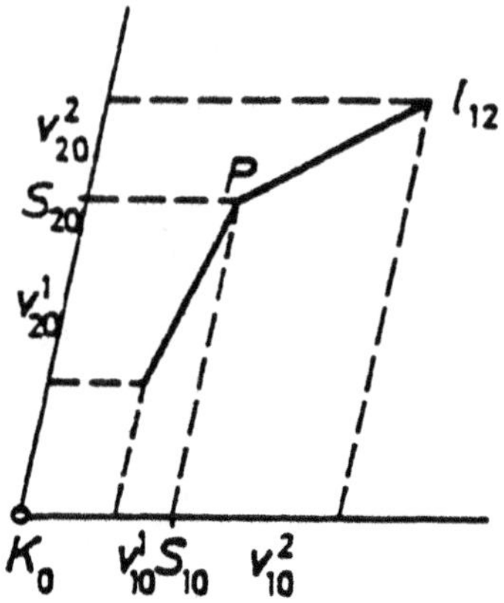

Fig. 14

K_3 with: $B_{30} \equiv B_{10}; b_{30} * b_{20}; K_1 \; r \; K_3, v_{20}^1$

K_4 with: B_{43} parallel to $l_{12}^1; b_{43} * b_{03}$;

K_5 with: $B_{50} \equiv B_{10}; b_{50} * b_{20}; K_1 \; r \; K_5, v_{20}^2$

K_6 with: B_{65} parallel to $l_{12}^2; b_{65} * b_{05}$.

If one regards the motions b_{13} and b_{15}, which body K_1 carries out in respect to the newly introduced reference bodies K_3 and K_5, then the positions s_{13} and s_{15} result as positions concomitant to s_{20}; s_{13} and s_{15} are assumed at places S_{13} and S_{15}. For the sake of better lucidity in Figure 15 the trajectories B_{43} and B_{65} are so drawn that they intersect at the places S_{43} and S_{65} belonging to P; to be sure, this represents no relevant supplementary condition.

The choice of conditions for the point-bodies K_3 to K_6 has the consequence that the guideline $l_{12;0}$, $l_{14;3}$ and $l_{16;5}$ are identical in segments (see Figure 15).

From the assumptions $v_{10}^2 * v_{20}^2$ and $b_{20} * b_{30}$ it follows by (S 1)c) and (S 5) that:

(I) $\qquad v_{13}^2 * v_{20}^2$.

From the assumptions $b_{20} * b_{30}$ and $b_{43} * b_{03}$ it follows by (S 1)c) and (S 2) that:

(II) $\qquad b_{20} * b_{43}$.

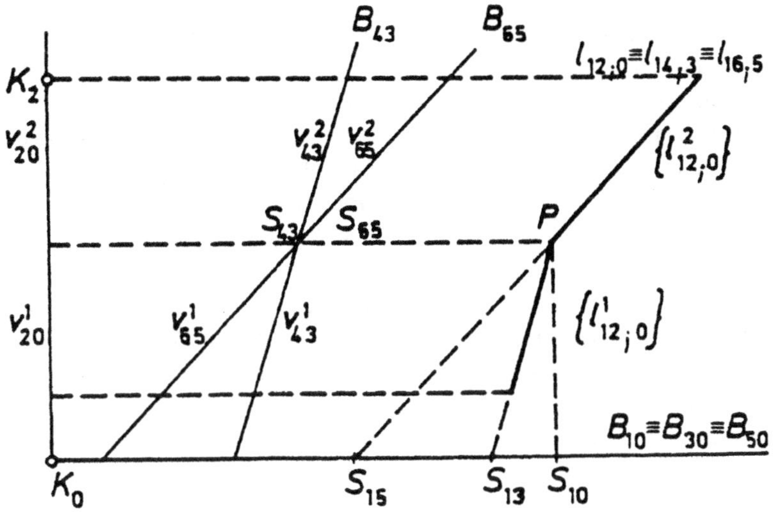

Fig. 15

From (I) and (II) follows:

(IIIa) $\qquad v^2_{13} * v^2_{43}$.

Analogously one proves:

(IIIb) $\qquad v^1_{15} * v^1_{65}$.

Therewith is shown: The auxiliary construction twice allows the application of definition (D 8), i.e., once to the point-bodies K_1, K_3 and K_4 and once to K_1, K_5, K_6. Then according to (D 8) the following holds:

$$K_1 \, r \, K_3, s_{43} \quad \text{and} \quad K_1 \, r \, K_5, s_{65}.$$

From (II) and $b_{20} * b_{65}$ – which is to be shown analogously – it follows for s_{20} by (S 6) that:

A motion b_{70} of a point-body K_7 can be produced for which holds:

$K_2 \, r \, K_0, s_{70}$ and by (S 6) then also

$K_1 \, r \, K_0, s_{70}$ or:

$K_2 \, r \, K_0, v_{70}$ and by (S 6) then also

$K_1 \, r \, K_0, v_{70}$ which was to be proven.

Theorem (S 7) now allows us to obtain propositions about the continuity of velocity changes without the use of clocks. For this, first, the state of affairs described by the assumptions of (S 7) must still be presented with other than the terms used here; by words, in particular, which make talk about velocities possible. In striving for the new description we will have to see to it that it agrees with practical usage to the extent that in clock-free kinematics, as well, the motion of a body is supposed to be called faster than that of another if it covers larger distances 'in equal times' (expressed by concomitant events). In the available terminology this may be defined as follows: The motion of a point-body K_1 is faster than the motion of a point-body K_2, if the path-length ratio $|w_{10}^n| : |w_{20}^n|$ of concomitant paths w_{10}^n and w_{20}^n is greater than l for all n. Extended to motion comparisons with arbitrary reference bodies this means:

A motion b_{1X} is faster than a motion b_{2Y} if for the path-length ratios of all concomitant paths the following holds:

$$|w_{1X}^n| : |w_{2Y}^n| > 1.$$

If one regards in Figure 10 the guideline $l_{12;0}$ and indicates the velocity ratios g^1 and g^2 for the pairs v_{10}^1, v_{20}^1 and v_{10}^2, v_{20}^2 of concomitant events by reference to their similarity by

$$g^1 = |w_{10}^1| : |w_{20}^1|,$$
$$g^2 = |w_{10}^2| : |w_{20}^2|,$$

then g^1 and g^2 are obviously different. Thus the velocity ratio of K_1 and K_2 changes at places S_{10} or S_{20}, and that is *erratically*, for which the term 'discontinuous' will be substituted.

Consequently the discontinuous change of velocity *ratios* is equivalent *per definitionem* to the occurrence of a bend-point in the guideline. Strictly speaking we must still show that this definitory equivalence can be extended to bends in non-straight guidelines, i.e., with non-similar partial events; however this may be omitted here for the sake of simplicity. But now, on the other hand, theorem (S 7) says that from a bent guideline one can infer that the bodies which produce the guideline assume rest positions in the places belonging to the point of the bend. From this logically follows: If two point-bodies assume no rest positions along a pair of concomitant paths, then they cannot change their velocity ratio discontinuously within these concomitant paths.

To be sure, with this we have obtained only a theorem about velocity ratios, while in the introductory discussion of the continuity problem the question was asked whether a point-body could discontinuously change its velocity without talking there about a comparison of velocity. However here it may be presuppossed as sufficiently established (see p. 95) that 'absolute' propositions about velocities at first can only be understood as propositions relative to the human time-sense and that, in the domain of the exact sciences, then, an objective time standard must take the place of this time-sense.

Since in this work the objective, i.e., reproducible, standard for comparing motions will be determined and a realization procedure given – since later a methodologically introduced talk about clocks will then be available – the comparison of arbitrary motions with a uniform motion of constant velocity can be included here in advance in the examination of continuity.

A clock-free, merely geometrical, motion comparison can provide no propositions about forms of motion. Thus also nothing is said by means of the course of the guideline about how the velocity of a body changes 'absolutely', and this can only mean here: relative to a clock. If one first disregards (S 7), then, from a discontinuous change of the velocity ratio of two bodies, one can only infer, for their 'absolute' changes of velocity, that either both bodies change their velocity discontinuously or, if one of the two bodies undergoes 'absolutely' only continuous changes of velocity, then at least the other body changes its velocity discontinuously. It thereby becomes clear that by ignoring (S 7) the introduction of clocks cannot decide for or against the possibility of discontinuous change in velocity. (To avoid a naively realistic misunderstanding it must be added that 'possibility' in this sentence merely means consistency with the already available theory.)

However, if one takes account of (S 7) and in advance refers to the workable terminology for motion forms which is still to be developed, then the result of (S 7) can be formulated as follows: An 'absolute' discontinuous velocity change, i.e., a velocity change which requires no duration, manifests itself through a bend in the guideline which occurs when a motion is compared with a 'clock-body', by which is meant a point-body which moves uniformly and with constant velocity. According to (S 7) this formulation would be equivalent to the statement that a uniform motion of constant velocity contains rest positions, something that is an obvious contradiction. If we exclude by definition the state of rest being a special form of a uniform motion of constant velocity, then, through the consideration of continuity, it is also shown for velocity ratios that there can be no discontinuous velocity change relative to a uniform motion; i.e., whoever agrees to the terminology common in classical physics, must also concede that a body can change its velocity only in a duration differing from zero.

From this insight a further consequence follows which shall be proven as the last theorem of clock-free kinematics:

(S 8) If a point-body describes a bent trajectory, then it has a rest position at the place of the bend.

For judicious reasons one can speak of bent trajectories only where at least three point-bodies are considered. The proof of (S 8) proceeds then from the following situation: Let a point-body K_1 move relative to K_0 along trajectory B_{10} (see Figure 16). Let trajectory B_{21} of a point-body K_2 be non-parallel to B_{10}. Let K_2 rest relative to K_1 and relative to v_{01} and otherwise move similarly to b_{10}. With the use of (S 3) a trajectory B_{201} (as abbreviation for 'the trajectory of K_2 relative to K_0 and K_1') results for K_2; B_{201} has a bend in $S_{20\dot{1}}$, the place concomitant to S_{10}.

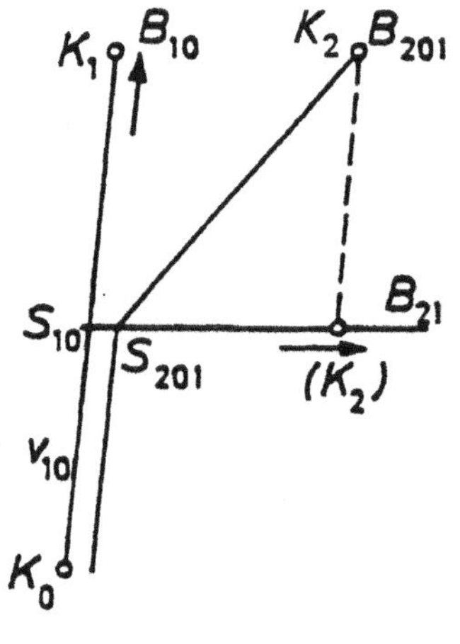

Fig. 16

If one now applies rule (R) to K_0 and K_1 (as it already is expressed in the assumption $K_2\, r\, K_1$, v_{01}, i.e., if one exchanges reference-bodies, then $l_{20;1}$ results as a new guideline (see Figure 17). This line exhibits a bend in P, the point belonging to $S_{01} = S_{10}$.

The course of $l_{20;1}$ is the same as that described in (D 8), so that the following holds: $K_2\, r\, K_1$, s_{01}, from which again by (R) it follows that K_2 has an (instantaneous) rest position at place S_{201} in its motion relative to K_0, which is what was to be proven.

Since relative rest, the rest of a point-body defined geometrically in accordance with (D 2) and (D 3), implies 'absolute' rest, i.e., rest relative to the motion of clocks – and clocks are supposed to assume no rest positions –, the 'absolute' validity of (S 8) may also be asserted. A body describing a bent trajectory assumes at the place of the bend a rest position relative to the motion of clocks. Thus in the 'ideal case' a body can describe a bent trajectory only if its velocity becomes zero at the place of the bend.

Extension of theorems (S 7) and (S 8) to non-similar motion can take place geometrically and offers in principle nothing kinematically new.

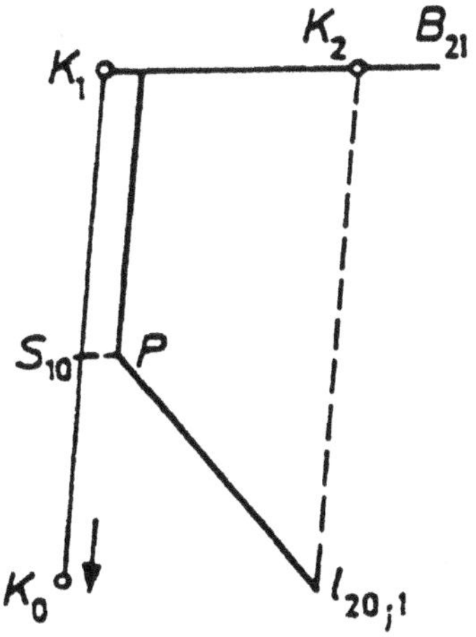

Fig. 17

Since in the following we will not have recourse to it, this extension shall not be given here. Thus with (S 8) the construction of a theory of motion that contains only propositions about simultaneous comparisons of motion is concluded. That it was developed only for motion in a straight line may be inferred from the goal of this work, namely of pointing out a way to introduce talk about time within the scope of the exact sciences.

Before we now turn to the introduction of terms for motion forms, we must still keep the promise given above. We must show that clock-free kinematics, even if it is not to be proposed for use in practice, nevertheless suffices to formulate nontrivial qualitative as well as quantitative empirical propositions as hypotheses that are accessible to experimental testing. For this purpose four examples of increasing complexity will be given in which only the terms available up to now occur as 'kinematic' predicates.

1. The free fall of a body is similar to the motion of a body along an inclined track when both motions begin simultaneously, i.e., when the starting positions of both motions are concomitant positions.

2. The motion of a body ascending on an inclined plane lags behind the motion of a body falling on an inclined plane.

The following theorem about falling motions goes back to Galileo:

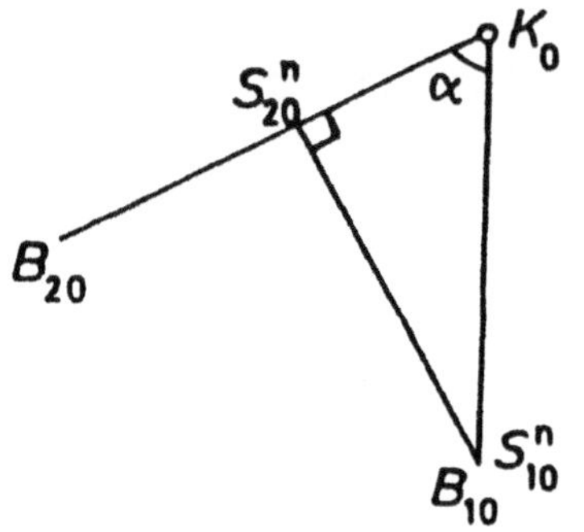

Fig. 18

3. Let bodies K_1 and K_2 fall relative to K_0, K_1 plumb, K_2 on an arbitrarily inclined plane. K_1 and K_2 assume concomitant positions at the interesction of their trajectories B_{10} and B_{20}. (This requirement may be fulfilled for extended bodies by projecting their trajectories onto a plane.) Then for all pairs of concomitant places the following holds: One can ascertain the place S_{20}^n concomitant to a place S_{10}^n by dropping a perpendicular from S_{10}^n to the trajectory B_{20} (see Figure 18).

Equivalent to this is the sentence: The ratio of lengths of the concomitant paths w_{20}^n and w_{10}^n is equal to the cosine of the angle which trajectories B_{10} and B_{20} form.

Finally the following, certainly no longer trivial, theorem can also be formulated without the use of clocks, a theorem which is empirically valid:

4. Let the trajectories of the bodies K_1 and K_2 form an angle α. Let B_{10} be plumb (see Figure 19). On B_{20} places S_{20}^1 and S_{20}^2 are specified by the following conditions:

1. $|w_{10}^1| : |w_{20}^1| = \cos \alpha$ (for S_{20}^1).

2. Starting from S a segment of w_{20}^1 of length $2\, w_{10}^1$ is marked off and the remaining segment is extended to the projection of w_{10}^1 onto B_{20}. The segment so obtained starting from S determines the place S_{20}^2 on B_{20}. Let the path form S_{20}^1 to S_{20}^2 be called w_{20}^d.

If then the following holds: $K_1 \, r \, K_0$, v^d_{20}; S^1_{10} is a place concomitant to S^2_{20} (i.e., K_1 begins to fall in the moment in which K_2 passes the place S^2_{20}); then the bodies K_1 and K_2 arrive at S simultaneously and have equal velocities at this place.

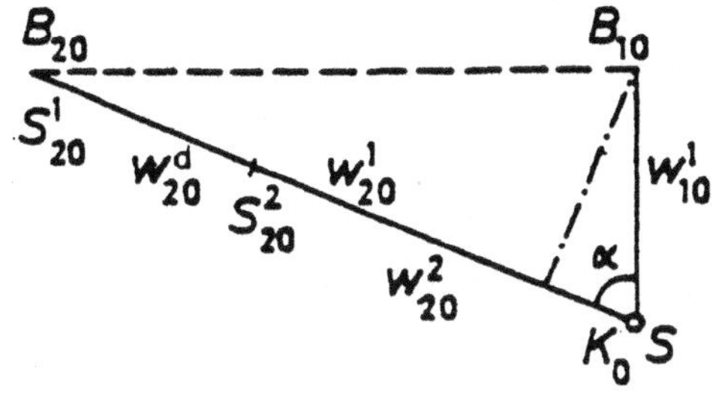

Fig. 19

4. FORMS OF MOTION

4.1. *Introduction: Objections to Periodicity as a Protophysical Basic Concept.*

If one measures what we have achieved up to now in clock-free kinematics against the program which was sketched in the discussion of the preliminary understanding of chronometry, then up to this point nothing more has been attained than the modest possibility of being able exactly to compare velocities in the case of simultaneous motions. We also spoke of this more in passing and only in respect to the proof of continuity, precisely because it may already be asserted that the problem of a theory of time-measurement does precisely not lie in the simultaneous comparison of motion.

For this reason we also refrained from additionally introducing a mode of speech about forms of motion, which first of all could contain only two-place predicates and which, therefore, – compared with ordinary usage – would have to remain at least an odd type of description. To be sure, by distinguishing definite forms of the guide-line, we can introduce predicates for 'relative forms of motion'; but terms such as 'accelerated relative to' (instead of 'outrun' in definition (D 5)) would be misleading. That is, a motion can be accelerated relative to another motion and likewise be 'absolutely', i.e., relative to a clock, retarded.

Speech about forms of motion only becomes meaningful when reference can be made without difficulties to a general standard motion and thus predicates for forms of motion are one-place predicates. That is also the case in ordinary language. Such a one-place predicate – it will be shown that one predicate already suffices to speak suitably about time – will now be introduced step by step; in doing so we will give a decision procedure which allows us to argue for or against the legitimate affirmation of this one-place predicate for a form of motion. Therewith, in particular, we will attain time-independence of propositions about forms of motion; this time-independence will mean that motions occurring at different times are comparable with respect to their form.

To the unbiased expert on elementary classical physics two forms of motion present themselves, namely 'periodic' and 'uniform' motion forms; one may expect of them that they may be designated as basic forms of a chronometry. (Here 'periodic' and 'uniform' are introduced first by example. In the following these predicates are used as contraries, i.e., a motion shall be called 'periodic' only if it is not uniform. Thus pendulum oscillations are an example of a periodic motion, but not the rotation of a body with constant angular velocity. Let 'uniform' designate only motions with constant velocity, but not uniformly accelerated motions.) Between both these forms we must now make a justifiable choice under the premises presented in Chapter II.

Without violating the principle of methodical order we can here tackle the proposal to be found in most physics textbooks which sees in periodic motion a form of motion enabling time-measurement. To maintain methodical order we must merely take care that the theorems of physics are not taken over uncritically. (The task of securing the conceptual foundation of empirical physics will thus obviously be understood here as first seeking out the methodical beginnings of theories recognized by physicists and, only when reasons against the possibility of a methodical foundation exhibit themselves, answering the deferred questions about presuppositions unhampered by recognized theories.)

The attempt to establish chronometry non-circularly requires a terminological definition of 'periodic'. That a definition via periodic functions of the time parameter t is inadmissible here is self-evident. Thus, as a rule, one goes back to formulations in pre- or non-scientific speech and tries to describe with them the indistinguishability of single periods. One speaks, frequently illustrated in examples such as the string pendulum, of 'states' or 'phases' that repeat themselves, such as, say, the return of a pendulum. Notable in such attempts is not only the fixation upon the old program in the theory of science to reduce measurement directly to the counting of identical objects, but also the bias for an unreflected practice of time measurement: because employing clocks with oscillators like pendulums, balance-wheels etc., was historically successful, one indeed thinks, in the attempts to define 'periodic', immediately of such examples of clocks and thus overlooks at least two problems. Firstly one tacitly goes through a pragmatic circle when with the help of clocks (in some sense or other confirmed) one defines a form of motion with whose help 'clock' is supposed to be defined. And secondly one passes over the task of separating relevant from irrelevant differences in the definition of 'periodic'. Clearly there must be such differences, otherwise one could not speak of different periods and count them for the purpose of chronometry. If one considers some other event at the same time as the start of the motion to be designated as periodic, then very many differences can be stated between two successive, and in some features indistinguishable, 'periods'. *De facto* without committing a circle, however, one can neither know that these are all relevant, nor that they are all irrelevant.

On the other hand, the possibility of at least precisely designating some features in which the periods of a motion should not differ, appears unproblematical. Although as a rule, use is not made of them in physics textbooks, all geometric propositions, for example, are available for this.

If now in the usual presentations there already exists an incriminating deficiency in regard to the determination of 'periodic', it remains fully obscure why the more vaguely described motions should be suitable for time-measurement, i.e., what connection exists between the 'definition' of 'periodic' and the de facto expected operating properties of clocks, namely their constant running ratio. Thus, if one maintains that features which are already thoroughly exact may be given non-circularly for the definition of 'periodic', namely e.g., geometric propositions about motions, then the core problem of this definition still lies in distinguishing relevant and irrelevant differences of individual 'periods'.

With what justification does one count, e.g., with a 'periodically' moving body, geometric criteria for the track among the relevant criteria? If, say, a string pendulum of constant length relative to the suspension circumscribes irregular tracks, while in the same geometric reference system, however, another string pendulum circumscribes regular ones, one will generally prefer for time-measurements the geometrically regularly swinging pendulum. This example shows how difficult it is to keep the definitions of physical basic concepts free from borrowing from physical knowledge that one, for his part, can only obtain with clocks. Moreover the example reveals a logical problem of these definitional attempts, a problem that has only been noted by a few theorists of science: assumed that one has agreed – for whatever reason – to regard a certain feature as relevant for the definition of 'periodic'. Then one infers: If differences between individual 'periods' in respect to this feature exist, then the motion in question is not periodic. However from this it does not logically follow that a motion with individual periods that are indistinguishable in this feature is periodic. Rather it can only be inferred that with periodic motions no differences in the given feature exist. In other words, and in the terminology of Chapter II, one is not falling back upon a definition, i.e., a definitional equivalence, but only upon a predicate rule, i.e., a definitional implication. The absence of the distinction of relevant from irrelevant features of periodic motions thus is expressed here in the fact that no definition, rather only a rule, presents itself; this rule states that periodic motions should not exhibit any difference in individual periods. A definition in the sense of a definitional equivalence only can be gained from this when a catalog of features, a catalog designated as complete, is given, relative to which features individual periods of a motion must be indistinguishable.

As already remarked in Chapter II, 2, it are not only empiricists who incline to the view that one could successively complete such a catalog of features in the progress of empirical physics, but also e.g., H. Dingler. One has to visualize this view, say, in such a manner that at first an apparently periodic motion is used for time-measurement. When, then, knowledge e.g., about the influence of air resistance on moving bodies has been gained therewith, the constancy of air resistance in all periods may presumably be added as a new feature to the definition and a clock can be improved in this direction. In this procedure there lies, however, an error insofar as, for detecting the braking effect of air resistance in dependence on e.g., temperature and air pressure, it must be assumed that the clock that is not yet stabilized relative to air resistance already runs constantly. One must, in other words, accept the contradictory argument that both the stabilized as well as the non-stabilized clock are good clocks, though both do not continuously exhibit a constant running ratio. This state of affairs may also be expressed by remarking that no reason can be given for taking the stabilized clock to be the better one, because there is no criterion of improvement that is independent from time-measurement. With that there only remains requiring the indistinguishability of periods in respect to all parameters that ever occur in physics. That is, however, obviously not a useful suggestion. Firstly, for parameters whose definition or measurement depends on time measurement, it contains a methodological circle. Secondly, one accepts, pointedly stated, a practice of time-measurement whose purpose one can determine completely only with the help of time-measurement itself. Thirdly, perhaps also parameters such as the age of the universe should be counted among the features to be taken into consideration here because of possible connections with gravitational constants – something which indeed yields no objections to an accurate measurement of time sufficient for practical purposes, but which does lead to a definition of 'periodic' according to which there can be no periodic motions at all in the strict sense.

R. Carnap already has seen these difficulties clearly from another angle and sought a way out. Carnap's criticism of attempts to define periodicity may be conceived as follows: If time is to be measured with the help of periodic motions, i.e., if the equality of durations is to be ascertained, then the equality of durations cannot belong to the definiens of 'periodic'. Therefore Carnap distinguishes 'weakly periodic' motions, in which some sort of phase continuously repeated itself, from 'strongly periodic' ones, in which the repeating motion phases last equally long. He demonstrates that the requirement of strong periodicity as a basic form of time measurement is circular and suggests, therefore, the following way out: One begins with weakly periodic motions. With their help one can discover that in nature there is a large number of weakly periodic motions which have a constant frequency ratio relative to each other. And so far as one chooses motions from this group for the foundation of time-measurement, the description of nature gained thereby would be simpler than e.g., with time-measurement with the help of our pulse.

That this view of Carnap of the nature of time-measurement is not even a correct description of what the physicist actually does and that it contains some completely nonsensical assumptions was already shown in Chapter I. Here Carnap's view is cited once more to extract its reasonable core. Agreeing with Carnap, it may be asserted that words like 'equaly long lasting' or their equivalents may not go into the definition of 'periodic'. Thus one must manage with a more sparing definition. But because we, and here a departure from Carnap is unavoidable, seek 'natural laws' whose validity is time invariant, we must artificially produce processes for time-measurement that are time-invariantly reproducible. Carnap's intention of additionally legitimizing reference to a class of, in his erroneous opinion, motions that 'exist by nature' by the fact that the simplest descriptions of nature would result from it, may be most easily interpreted with the consistency thesis of v. Weizsäcker, 1971, p. 196. According to this thesis the distinguishing of measurement processes must be logically compatible with the empirical theories obtained therefrom. Within the scope of such a consistency consideration one can derive then from the empirical physical theory reasons for why the at first weakly periodic motions are equally long in their periods. Such an interpretation, probably in line with Carnap's intentions, reveals that Carnap indeed succeeded in avoiding a blatant definitional circle, but that for doing that he runs through an even more incriminating methodical circle.

Can this methodical circle be avoided? Is, in other words, a protophysical definition of periodic motion possible?

Since in this place clock-free kinematics is already available, non-circular statements (in means thus richer than in Carnap) can be made not only about ('naturally') distinguished states such as the beginning and end of processes, but also about motion comparisons for arbitrary internal phases of motions. Thus e.g., the motions of two synchronously oscillating pendulums (also with different amplitude) are similar to each other. Admittedly the definition of 'period' shall serve precisely to compare non-simultaneous durations. With that it appears first questionable whether the vocabulary defined for simultaneous motions can at all be applied to non-simultaneous motions. Obviously this will not be possible by, say, requiring that two successive processes should be similar. Such a requirement is, in accord with the definition of 'similar', meaningless. Therefore we shall now discuss the attempt to formulate a definition of 'periodic' in connection with the comparison of simultaneous motions.

In clock-free kinematics the possibility of a momentary (simultaneous) velocity comparison was obtained. If one now takes into consideration the momentary relative velocity of two motions b_1 and b_2 as a function of the distance travelled by one of the two motions, e.g., by b_1, then a *relative periodicity* may be defined by the fact that this function is periodic. It is then a mere geometric or mathematical problem to prove that the choice of one of the two motions plays no role as a reference for the velocity ratio, and that all motions that are periodic relatively to each other form an equivalence class with respect to relative periodicity. In the same way it is unimportant with what relative phase shift both motions b_1 and b_2 start. Obviously, therewith a rigorous definition of a class of motions has been found that corresponds to the, according to Carnap's view, naturally existing class of periodic motions. Thereby it plays no role that the definition given here obviously has advantages over Carnap's, insofar as it is independent of phase shifts and does not have to presuppose that the frequency ratio of both motions is rational. If relative periodicity is defined the definition of periodic motion ('with periodic' as a one-place-predicate) and of a clock can be obtained in a way quite analogous to that which is based on the concept of similar motions and which is shown in Section 4.2 of this chapter. We have to postulate for the motions belonging to one and the same class of relatively periodic motions that the relative periodicity is maintained if the phases or starting points of two motions are shifted relative to each other. For short, we shall say that those motions are relatively shiftable. Then a motion is defined as 'periodic' (one-place-predicate) if it belongs to a class of relatively periodic and shiftable motions.

Although periodicity can be defined with protophysical rigor good reasons stand against this way:

Firstly the definitions and especially the following theorems, particularly the theorem of uniqueness and its proof, are much more complicated than in the case of similar motions. The constant factor characterizing the ratio of two velocities has to be replaced by a periodic function which may become very complicated if the two motions in comparison are shifted relative to each other. A short glance at the proof of the uniqueness of the uniform motion shows that a product of three periodic functions would have to be formed, functions about which nothing is known but their periodicity. However, this first argument against periodicity as defining form of motion for clocks only concerns mathematical complexity.

The second argument against periodicity is based on epistemological reasons. Throughout history clockmakers tried to build clocks the hands of which simulate the motion of the sun around the earth or, for short, the earth rotation. Periodic motions turned out to be the best technical means in order to reach this goal in mechanical clocks. Periodic motions, say, of a pendulum were the necessary eval in order to retard the velocity of a falling weight which was directly indicated by the clockhand. Only later the falling weight was interpreted in a new way as the driving force of the clock which should compensate the friction losses of the oscillating body, and the clockhand was interpreted as a mere counting device of the number of the repeated periods of the oscillating motion. This 'new' interpretation which is the common view in modern physics textbooks presupposes a knowledge of the laws of certain oscillators like pendulums, that is to say, of laws accessible only by means of clocks or at least by means of an operational concept of time.

To choose periodic motion as the definitoric form of motion for clocks therefore entails the epistemological error of a methodical circle. There is no natural law or logical need to persue the goal of time measurement, namely the quantitative comparison of two durations, by reproduction of time units or of events of the same duration. As shown in Chapter I the reduction of measurement to the counting of units is an error arising from a particular empiristic tradition which forgot the pragmatic conditions of the reproduction of units.

A third counterargument against periodic motion concerns the fact that uniform motions as a possible alternative are usually regarded only as inertial motions the definition of which indeed presupposes time-measurement. Section 4.2 however shows that uniform motion can be defined merely by means of clockfree kinematics and without any dynamic terms so that from the problems about inertial mechanics no reasons emerge in favour of periodic motions.

4.2. *Uniform Motion.*

If we now embark upon the task of defining a one-place predicate for forms of motion, we must start from unquestionable prescientific practice. The construction of chronometry must already start, in particular, with talk about motions of bodies which is presupposed to be available insofar as speech about the repetition of events is meaningful. Non-scientific ordinary language always actually reaches such a stage wherever asking a person to repeat or neglect an action belongs to the most elementary situations. Even in the domain of 'externally observable' motions, i.e., such changes in which obviously no person cooperates, non-scientific language accomplishes the required identification. Well-known examples are the course of the sun, the surf or, say, the yearly flood of the Nile, which historically have played an important role for chronology. Even in the case of such naturally repeated events, reference to the situation of the man who acts purposively is crucial. It is he who identifies events and their repetition insofar as it meets his needs.

Since, in historical physics, the question occurs explicitly in many places (even outside of quantum physics) whether and under what conditions natural as well as artificial events are repeatable, one may easily fear that starting out from the repeatability of motions in the prescientific domain is at least a weighty decision with possible later unwanted consequences for physical theories. This apprehension is reason enough to once more critically inquire about the justification of this first step. The mere possibility of distinguishing repeatable events on the prescientific level and therein of treating chronometry analogously to geometry has already been shown. The goal of the endeavors we are engaged in is time-measurement. If one tries to put oneself at a stage where no clocks, not even the most primitive ones, were yet available and asks how it may well have been that a need for chronometers could have arisen (independently of how it happened historically), then the unreliable judgment of durations according to the time-sense would have been occasion for improving the estimation of durations. That indeed the time-sense occasionally leads to error cannot be ascertained or asserted without contrasting some sort of repeated event with the subjective judgment of a duration. But what is then expected of a chronometer (above all for everyday use and without any scientific claim) is nothing other than that it exhibits equal repetitions of a process in a manner which is at all times prescribable; 'equal' is here a pleonasm, since repetition will only mean what is again recognizable. Thus without already speaking of repetitions of events, not once yet did we have to examine the need for 'time-comparison' or 'time-determination' on the most elementary level of daily life, to say nothing at all about the suitability of a device for this purpose.

Here an argument still appears to present itself for the proposal which was rejected above, namely of having recourse to periodic motions for the foundation of chronometry. However this is not the case. The prescientific identification of repeatable processes has a much more elementary character than the definition of periodicity. The role of the prescientific identification of events may perhaps be illustrated by a geometric analogy: Just as an operative geometry cannot succeed by manipulating very soft bodies (say of wax or India-rubber), so it is obvious, prior to any science, that chronometry cannot be based on irretrievable events such as the movement of clouds, snow storms or even movements of man and animal. The downward rolling of a ball via an inclined chute, the oscillations of a string pendulum, the emptying of a receptacle through a small opening, etc. are, in contrast, equivalent to the selection of basalt or steel in the production of planar disks. Against such a choice of 'suitable' motions that are already accessible to everyday experience objections are to be expected. Let it be premised that in modern physics, as is well-known, not only laws of motion but also geometry is considered as a domain of basic concepts that as a result of an analysis of recognized physical theories is dependent on their empirical verification, that this domain, in short, is empirical. In classical physics, however, Euclidean geometry is presupposed naively. Propositions about motions, however, whether they concern chronometry or whether they concern kinematics and dynamics, are also considered, already in classical physics, as parts of an empirical mechanics. In particular there are no attempts to develop a theory of motion really relevant for the physicist independently of dynamical arguments (namely references to inertial motions).

Kinematics is understood as dynamics with questions of moving forces played down.

Considered from this viewpoint, beginning with repeatable motions on the basis of everyday knowledge appears possibly problematic. Stated pointedly, according to this view everything that moves, moves according to natural laws, because already the simplest propositions about the moving object are parts of an empirical theory, i.e., of a theory formulating natural laws. Replies to this possible objection have already been given in another context and proved there: Every repitition of equal events that is considered as naturally lawful, in order for it to be at all knowable, requires the human ability to make distinctions such that he can argue for or against assertions that an event repeats itself. In general, everyday experience cannot be circumvented by scientific experience. One would thus only gain a pseudo-certainty should one wish to bring into play physical reasons for the selection of first repeatable motions, i.e., for pendulum or inertial motions. Only selection criteria independent from the empirical results of physics are methodologically admissible. What these could be in detail will be discussed more thoroughly later.

Firstly it suffices to begin with motions of bodies that may be repeated. The expression 'repeat' is thereby defined exemplarily. Examples would be, say: throwing or dropping a stone, turning a potter's wheel, pouring a pitcher of water into another pitcher. Counterexamples are, say: igniting a (certain) match, drinking a particular bottle of wine empty, breaking a particular pane of glass. Already in such examples and counterexamples it can be illustrated then that by 'repetition' is not meant the general schema of an action (e.g., the igniting of a match), but rather the action schema together with particular bodies designated by proper names or definite descriptions (to 'ignite this match'). Then each description concerning the motion of a particular body can be defined exactly step-by-step, whereby each feature of the description can be verified according to whether the repeatability of the action is lost or not. In this manner one can learn what it means to repeat a motion that has been specified in an enumeration of properties and then to speak generally of a type of repeatable motions, say of the projectile motions of stones.

Likewise one can also learn, without a physical vocabulary, when a motion is left to itself, namely when it is no longer influenced by a man. The moving of a chess piece is not a motion left to itself, though the rolling of a ball in bowling or the flight of an arrow is. Motions left to themselves thus form a group, so to speak, that stands between motions guided completely by man and motions that run completely without human influence, such as the course of the sun or meteorological phenomena.

Obviously many sophistical objections to this division may be constructed. Assume that a psychologist, for an experiment in learning theory, has furnished a bowling alley with a contrivance in order to influence the course of the ball in a way unnoticed by the experimental subject. Or someone believes himself to have repeatedly caused a particular motion and then to have left it to itself; and instead of this motion would have started even without his influence and the presumed agent of the motion was led unnoticed to his actions – which in truth caused nothing – only by a by-product of this motion, etc. Such 'what would happen, if' questions miss, however, the intention of dividing motions into guided ones, those left to themselves and those not caused by human influence. There it is a matter of the question of what motions men can so set in operation that one can speak meaningfully of repeatability. And it is also a matter of finding examples for motions left to themselves which, although from a certain point in time on, are no longer subject to the direct influence of their human originator, nevertheless in their course, satisfy certain expectations of the originator. Every leap over a ditch, every throw or shot at a target is such an example. And again it is unthinkable in light of the methodological sequence of steps that men would have arrived at scientifically describing, explaining and predicting projectile motions, if aimed throwing did not belong to the elementary everyday abilities of man.

Clock-free kinematics, which was constructible solely with guided motions, may now also be employed to refine propositions about motions left to themselves and thus permit one to speak of repeatability in a more rigorous sense. With this a transition to the construction of instruments immediately ensues. Examples of motions that are repeatable in the stricter sense are, say, the courses of spring or weight driven toys. Even here the far-reaching parallelism of a methodological sequence and actual historical development again manifests itself: the construction of music automata and play clocks was not, say, a by-product of a watchmaker's trade directed at chronometry; rather the ability to establish increasingly complex mechanical (and acoustical) processes instrumentally also had the special goal of causing imitation of the earth's rotation by an instrument to be better attainable.

After these prefatory remarks about the manual, prescientific production of instruments which can be set in operation and then execute motions left to themselves, we shall now again attempt to come closer to the definition of uniform motion with the vocabulary of clock-free kinematics alone. For this purpose let the point bodies K_1 and K_2 on two instruments G_1 and G_2

be considered. K_1 or K_2 are thus parts of G_1 or G_2. However, otherwise the mechanism of G_1 and G_2 is constituted, we shall now be interested only in the motion of K_1 or K_2. K_1 and K_2 are, in other words, the 'pointers' of the instruments. Let the reference bodies for the pointer motions, i.e., the dials as it were (although here particular scales are not yet to be though of) be fixed for each instrument. In the following they will not be mentioned further nor will they be taken account of as indices, something that is proven to be admissible by the results of clock-free kinematics. With such instruments, now, clock-free, i.e., simultaneous, comparisons of motion will be carried out. Thereby the guideline (or the momentary velocity ratio in dependence on one arbitrary motion of the two) would be respectively registered.

(D 9) If repetitions of the motion comparison are begun respectively with the same pair of concomitant positions and if there arises thereby the same guideline, then the compared events may be called '*repeatable relative to one another*'. If the operations of two instruments G_1 and G_2, i.e., the events of their pointers K_1 and K_2 are repeatable relative to one another, they may be called '*invariable relative to one another*'. The establishment of a chronometry in the narrower sense, i.e., disregarding its presuppositions in geometry and clock-free kinematics, begins with the norm:

(N 1) Produce instruments that are invariable relative to one another! Equivalent to this is the norm: Produce instruments for originating events that are repeatable relative to one another! These norms are selection norms, i.e., they prescribe selecting from the domain of available artificial motions those motions for which relative repeatability is given.

It may be surprising that a norm that, after all, concerns artificially produced instruments is designated as a selection norm and not as a production norm. A closer examination of the methodological sequence of individual steps shows, however, that in the first step no rules for the production of invariable instruments can be formulated, because for this the knowledge is already requisite of what circumstances give rise to the invariability of instruments or what construction principles such a production must follow. In other words it must be presupposed that the production of events repeatable relative to one another has already succeeded.

Here it could be objected from the empirical standpoint that experience is also the basis of a normatively interpreted chronometry. Whether this view is to be accepted or denied depends, however, completely on the use of the word 'experience' in this sentence. It would be absurd to understand by this word an experimental result of a systematic endeavor and to thereby wish to ground the success of the selection of repeatable events by reference to a 'natural law'. That certain activities such as the production of invariable instruments are successful is, however, a quite empirical knowledge acquired in action proper (without guidance by principles as they are constitutive for the sciences). One could, apart from the distinction' scientific-prescientific' or 'measuring-nonmeasuring', also characterize this by the distinction 'historical experience-systematic experience'. The historical experience that we are successful in constructing instruments that always operate in the same manner (in the sense of relative repeatability) can, as it were, be had only once. It is then available and even later cannot be denied or taken back by a systematic experience within the framework of a physics working with repeatable experiments. Historical experience is, namely, unique in the sense that each new experience apparently contradicting it is acquired under new conditions and thus also concerns a new subject. With the assertion that a normatively established chronometry begins with a selection norm concerning the domain of already artificially produced instruments or motions, is thus connected an historical fact that, for its part, naturally again stands open to scientific questions but which is no longer a subject of physics.

It is a characteristic of selection norms that they cannot be satisfied increasingly better by approximation (as will be the case later with production and use norms), and that in the following sense: To be sure the 'observational accuracy' can be increased for the repeatability of events and it can also be the case that thereby some events, first considered to be repeatable no longer prove to be repeatable under more accurate controls. But there can be no procedure for improving repeatability, and that for methodological reasons. Events are either repeatable or they are not, whereby the respective underlying degree of observational accuracy is given by the technical ends for which repeatable events are employed. Thus, whether or not events of a particular type may be taken as repeatable is decided only by technical ends and not, say, by physical laws (e.g., by the relationship to the velocity of light in a vacuum).

To the repeatability of events there falls in chronometry a role analagous to the one for the (prescientificially determinable) hardness of bodies in geometry. There one first distinguished, by way of examples, between solid, liquid and gaseous. This distinction is laden with a large number of vaguenesses. It is thus defined exactly according to the respective purpose for which it will be used. In connection with the production of particular instruments and shapes of bodies the selection norm will be formulated that only 'hard' bodies be used for the production of instruments, where 'hardness' is defined as the material property of admitting 'identical in shape' (e.g., two bronze statues poured in the same mould are 'identical in shape') as a transitive relation. Thus as hardness (i.e., a relative constancy of form) is a selection criterion for materials with which one can first meaningfully begin to work bodies in accord with particular processes and to establish, say, norms for the production of measuring instruments, so events repeatable relative to one another are, as it were, the material with which one can first meaningfully begin to search out a particular form of events that is reproducible any time.

In analogy to similar objections to a protophysical foundation of geometry we shall already here take up the question of whether there is *only a single class of events repeatable relative to one another* or whether several classes could not be found, which would possibly result in several physics with a respectively different chronometry. The theories of time-measurement which start with periodic motions also deal with a similar question. There, then, as a rule, it is asserted, and with a certain relief at the burdens of proof seemingly satisfied thereby, that one knows from experience that there is only one large class of such periodic motions. In Chapter I, it was pointed out that this assertion is plainly false and that finding a large number of (even 'naturally') given classes of periodic motions that have no constant frequency ratio from class to class depends only on particular preliminary decisions in regard to the use of time-measurements. The argument given there holds analogously here: Already the question of whether several or only one class of events repeatable relative to one another are 'given' starts from an unacceptable fiction. It cannot be the goal of modern physics to comprehend the totality of nature and to ask thereby whether nature itself makes available for use only one or several ways for doing this. The possibility that several 'chronometries' could be found ('accidentally' or 'in accord with natural law') is, disquieting however, for the uniqueness and assertory force of physical theories only with this digressive premise of seeking to totally explain nature. The fear that, in nature, there could possibly be several classes of motions that are suitable for time measurement, on the contrary, does not arise within the framework of a consciously goal-oriented physics. For the case that, relative to a particular end, actually several classes of motions periodic relative to each other or several classes of repeatable events should be found, there remains still the selection of a single class according to the criterion of which of these classes is the most appropriate means in light of these ends. What admittedly are or should be the ends of physics or parts of physics cannot be learned from an analysis of its theories.

We promised above still more detailed instructions for the *selection of repeatable motions* independently of the results of empirical physics. We can now make this promise good. Starting from the insight that 'time' within the framework of the natural sciences is not a proper entity like, say, the earth or its field of gravity, but rather only an economical manner of speaking about the comparison of events, we must decide, in regard to chronometry, for what events a standardized manner of speaking shall be established. Thus it is e.g., trivial that for the purpose of determining distances on the sea a 'time-measurement' is needed which solely has the task of simulating earth rotation. Due to the great significance of the solar day, this chronometry is suitable for our daily life as well as for nearly all other tasks of chronology, particularly for organizational ones. It is obviously not suited, however, for all physical or general tasks in the natural sciences. Also, choosing for astronomical observations a 'clock' that likewise imitates the rotation of the earth (say for the purpose of subdividing the sidereal day) is certainly to be recognized immediately to be meaningful. If, on the contrary, times are to be measured in order to establish and test a theory of falling and projectile motions, the problem of selecting a type of events for chronometry arises anew. It can remain open whether it was an historical fortune or an intuitively correct anticipation of the inertial principle by the fathers of classical physics that pendulum clocks were employed, and with this a class of falling or projectile motions already chosen; or appealing to Newton's fusion of earthly and celestial mechanics whether mechanics and gravitational theory was intended as a theory of inertial motions and for that reason chronometry was carried out with inertial motions. So considered, electrodynamics, as long as it concerns itself with force measurements on moving bodies, must hold itself to time-measurement with the help of moving bodies; where, however, it investigates e.g., propagation phenomena of electromagnetic waves it must enter anew into a discussion of the selection of a class of events that measure time in this new context. Whether both these classes, mechanical and electrodynamic chronometric events are to be claimed as a single class, because 'clocks' of both types have a constant running ratio relative to each other is then a question to be left to measurement physics, a question whose answer, to be sure, however it turns out, cannot be a 'refutation' or 'revision' of a mechanical chronometry by an electromagnetic one. The presuppositions made by choosing repeatable motions to be the starting point are in part stronger, in part weaker than those of traditional theories of chronometry. Stronger insofar as they already prephysically take into consideration selection criteria for classes of artifical motions with the help of research goals, weaker insofar as they contain no hypotheses about a nature that provides us precisely with a class of periodic motions that have a constant frequency relative to each other. That, however, men can learn and perform actions that are repeatable in respect to designatable criteria is a presupposition that no one can dispute if he does not wish at the same time to dispute the possibility of any science at all (or other cultural disciplines).

Now after instruments invariable relative to one another, and motions repeatable relative to one another, are available, further norms will be formulated. For this we first define:

(D 10) Let two instruments G_1 and G_2 be invariable relative to one another. If the running ratio of G_1 and G_2 is constant, i.e., if the events of their pointers K_1 and K_2 are similar, then G_1 and G_2 '*agree with each other*', the events of their indicators are '*repeatably similar*'. With that, by (D 9), the velocity rations are equal with repetition. In definition (D 10) it was presupposed, in accord with (D 9), that with repetitions of the motion comparison there exists respectively the same ordering of concomitant positions for the beginning of the compared events. This restriction shall be abandoned in the following definition:

(D 11) Let two instruments G_1 and G_2 agree with each other. If then with repetition of the operations without maintaining identical starting positions a likewise equal and constant running ratio exists, i.e., if arbitrary partial events of the pointers K_1 and K_2 are similar to each other and have an equal momentary velocity ratio, then G_1 and G_2 may be called '*equal to each other*' and the events of their pointers '*displaceable relative to one another*'.

With this now, norms can be formulated as follows:

(N 2) Produce instruments that agree with each other!

(N 3) Produce instruments that are equal to each other!

(N 2) and (N 3) are, in contrast to (N 1) *production norms*.

Namely one can attempt gradually to change invariable instruments so that they agree and are equal. Let one think, say, of a pair of balls each one of which rolls down a chute. Such instruments are invariable to each other (independently from the geometric form of the chute). Assume that one chute is a straight line, the other in the projection on the horizontal plane is straight, in the projection on the vertical plane is concavely circular. Then these instruments do not agree with each other. It may, however, be ascertained by a clock-free motion comparison what the deviations from agreement are, more precisely that the straight moving ball outruns the circularly moving ball in the sense of definition (D 7), in case both balls start 'at the same time', i.e., in case their starting positions are concomitant positions. Moreover in this motion comparison it can be ascertained that raising the incline of one of the two chutes indeed raises the momentary velocity ratio to the motion on the other chute, but does not yet lead to agreement. From this the production instruction can already be obtained to eliminate the advance of the circularly moving ball by increasingly raising the incline of the circular trajectory along with the path. This way which naturally is artificially complicated only for illustration, leads finally to the choice of two straight chutes. They represent, then, two instruments that indeed agree with each other, but that are not equal to each other. For if one ball starts 'earlier' than the other, their motions are not similar (a state of affairs which will be described later in clock-dependent language, i.e., in the scope of a theory of falling and projectile motions, by the assertion that the velocity of falling motions in the field of gravity increases proportionately to time).

Again the clock-free comparison of motion can give rise to changes in both fall apparatuses so that finally fall apparatuses equal to each other will be obtained. For 'empirical knowledge' it suffices also here again to know that raising the incline of one of the falling motions leads to a raising of the velocity ratio relative to the other falling motion. The fall apparatuses are then equal to each other when (as will be said here, in anticipation of a terminology that is methodologically available only later,) the frictional losses of the rolling motion are compensated for exactly by the acceleration of gravity. Such motions may be realized (roughly and naturally unsuitable for accurate time-measurement) by metal cylinders that roll down inclined planes covered with velvet.

A hitherto implicit presupposition must yet be explicitly stated: With all attempts to contruct a chronometry upon repeatable motions, only finite motions, i.e., terminologically precisely – events, come into consideration, because 'eternal' motions cannot be repeated *per definitionem*. Norm (N 3), which requires the displaceability of pointer events, however, precludes the fact that the total operations of instruments will satisfy this norm from rest position to rest position; for in some vicinity or other of the beginning and end position an acceleration or retardation of the indicator must of course ensue, otherwise the indicators would be always or never moving. Thus all assertions and norms made above always hold, except for a starting and stopping partial event.

It was our aim of the efforts hitherto undertaken to find a non-circular definition of uniform motion. A glance at norm (N 3) shows that here (in the following always disregarding a starting and stopping partial event) a *homogeneity of events* is required. The events of G_1 satisfying norm (N 3) are homogeneous insofar as all their positions are indistinguishable. With this, distinctions can be formulated and must in such a case, be ascertained in respect to some position or other of the pointer K_2 of an instrument G_2 that is equal to G_1. The sole criterion that thereby comes into question is the momentary velocity ratio ascertainable without clocks. It is thus precisely the requirement of equality of the instruments to be produced or of the displaceability of their pointer events that entails the homogeneity of these events.

In light of the illustrated homogeneity of events displaceable relative to each other, they shall also be called *'regular'*. *Regular* is thereby a predicate, determined through definition, for real events produced on particular instruments. Corresponding to the methodological illustrations of Chapter II, an ideation process must now still be given. For this purpose the chronometric homogeneity principle is established.

$$\text{(H)} \qquad S \in V \wedge S' \in V \wedge a(S, V) \rightarrow a(S', V)$$

i.e., two positions S and S' from the event V (in instrument G) are indistinguishable. Thereby 'a' are propositions in which no free variables can occur, though those bound by quantifiers may. By a replacement, admissible in this sense, of 'a' by a proposition that is possible with help of the vocabulary of clock-free kinematics, there arises from the homogeneity principle H, the homogeneity sentence H_R.

If one then restricts himself to those propositions A that logically follow from H, i.e., if one speaks about regular events invariantly in respect to realization deficiencies, then to terminologically characterize this transition, the ideator *'uniform'* shall take the place of 'regular'.

For the sake of terminological completeness a supplementation is called for: A look at the individual steps in the introduction of 'uniform' shows that this term was first defined only for events. Nevertheless its application to motions is easily assured *per definitionem*: Let a motion which is uniform in all its partial events be called uniform. If a motion is uniform, it thus has no rest positions, which can also be expressed thusly: Uniform motion is endless.

(D 12) An instrument on which a point-body (pointer) moves uniformly is called a *clock*.

As pointed out in other places, the question of a time metric, here i.e., of a clock scale, shall be transferred to a geometry; moreover questions of the definition of a time unit within protophysics also are of no meaning.

Here it is only a question of norms for the functioning of instruments that ideatively define the uniformity of pointer motions in clocks. Therewith the definition of clock pursues the goal of regulating the production and use of clocks and of distinguishing undisturbed from disturbed operations.

4.3. *The Uniqueness of Clock Definition.*

According to Chapter II, the uniqueness of homogenous uniform motion has now still to be proved. To be able to lucidly carry out this proof, the terminology shall be extended minimally. The (momentary) *velocity ratio of two concomitant positions* is determined by the slope of the tangent to the guideline at the point belonging to these positions. The slope can be given by the ratio of the lengths of the paths of two concomitant events that have this tangent to the guideline. In this way, i.e., by the formation of a limit, velocity ratios of positions may be reduced to ratios of path-lengths and represented by rational numbers.

With this then results directly a theorem for all quadruples of concomitant positions of motions b_x, b_y, b_u, and b_v, whose proof shall be omitted here due to its simplicity:

(S 9) $\quad g_{xy} = g_{xu} \cdot g_{uv} \cdot g_{vy}$.

The expectation of real clocks, i.e., of realizations of instruments with uniformly moving indicators, is, as paradoxical as it sounds, nothing other than that they be clocks in the sense of the definition; the uniqueness of this definition implies materially that clocks exhibit among themselves a constant velocity ratio and that even when they result from different and mutually independent realization procedures. Consequently in the uniqueness proof it is to be demonstrated that from the presupposition that two pairs of instruments G_1, G_2 and G_3, G_4 are respectively clocks it logically follows that each clock of one pair has a constant running speed ratio to each clock of the other pair, or, in the terminology introduced above that each clock agrees with and is equal to each other.

Thus in the sense of (D 9) and (D 12) let two pairs of clocks U_1, U_2 and U_3, U_4 be given. Without restricting general validity it shall then be shown that the pair U_1, U_3 mutually agrees and is equal. For the other three pairs the proofs run analogously.

Of the presuppositions, one needs explicitly the displaceability of the clock pairs U_1, U_2 and U_3, U_4:

$$(1) \qquad g(s_1^1\ s_2^1) = g(s_1^1\ s_2^2)^W;$$

$$(2) \qquad g(s_3^1\ s_4^1) = g(s_3^1\ s_4^2)^W;$$

The lower indices indicate the respective clock. The raised index 'W' indicates that on the right-hand side of each equation, it is a matter of velocity ratios that were ascertained in a repetition of those events from which the velocity ratios of the left-hand sides of the equations originate.

Moreover, by supposition, all four clocks are invariable relative to each other, i.e., there holds in particular:

$$(3) \qquad g(s_2^1\ s_4^1) = g(s_2^1\ s_4^1)^W;$$

$$(4) \qquad g(s_1^2\ s_3^2) = g(s_1^2\ s_1^2)^W;$$

$$(5) \qquad g(s_1^1\ s_2^1) = g(s_1^1\ s_2^1)^W.$$

Now according to theorem (S 9) there holds:

$$(6) \qquad g(s_1^1\ s_3^1) = g(s_1^1\ s_2^1) \cdot g(s_2^1\ s_4^1) \cdot g(s_4^1\ s_3^1);$$

and likewise according to (S 9):

$$(7) \qquad g(s_1^2\ s_3^2)^W = g(s_1^2\ s_2^1)^W \cdot g(s_2^1\ s_4^1)^W \cdot g(s_4^1\ s_3^2)^W.$$

If one compares the factors on the right-hand sides of (6) and (7) in order, the first are equal by (1), the second by (3), and the third by (2). With this is taken into consideration that displacement is a symmetrical relation. From the identity of the right-hand sides follows that of the left-hand sides:

$$g(s_1^1\ s_3^1) = g(s_1^2\ s_3^2)^W, \text{ where by (4),}$$

$$(8) \qquad g(s_1^1\ s_3^1) = g(s_1^2\ s_3^2) \text{ also holds.}$$

If one considers how in (1) and (2) the selection of the upper indices 1 and 2 takes place, then one sees immediately that equation (8) holds for all 1, 2 pairs of concomitant position pairs $(s_1\ s_3)$ of clocks U_1 and U_3, i.e., it has been proved that U_1 and U_3 agree with each other.

Now it remains to be shown that they are also displaceable relative to each other.

According to (S 9) there holds:

$$(9) \qquad g(s_1^1\ s_3^2)^W = g(s_1^1\ s_2^1)^W \cdot g(s_2^1\ s_4^1)^W \cdot g(s_4^1\ s_3^2)^W.$$

As above, a comparison of factors of the right-hand sides of equations (6) and (9) is performed. The factors are equal by (5) as well as by (3) and (2). Thus there also holds:

(10) $g(s_1^1\ s_3^1) = g(s_1^1\ s_3^2)^W$;

i.e., clocks U_1 and U_3 are displaceable relative to each other, Q.E.D. (Assumed, in addition, was that no four of the clocks is standing, i.e., that no $g()$ is zero or infinite.)

With this the uniqueness proof for chronometry has been given. Thus if two clocks from a class of instruments invariable relative to each other have no constant running ratio, this can only be due to a defective observation of norms (N 2) and (N 3) or to external disturbing influences. Such disturbances can then be eliminated by exhaustion.

Although it is no longer a part of protophysics to consider the course of the exhaustion procedure for empirically detecting disturbing causes, we shall illustrate here by example how, on the basis of the indicated norms in practice, the improvement of clocks should take place in principle. Such an example is, among other things, helpful because it has been asserted by some critics of protophysics (to be sure with a methodologically circular recourse to special relativity theory) that in the case of an observed deviation from the protophysical norms (N 2) and (N 3) there is in principle no possibility of eliminating disturbances, so that finally the claim to a uniquely grounded chronometry also breaks down. (So for instance W. Büchel, 1970, replied to by Janich, 1976b). Thus, first, an example shall be presented and then once more the uniqueness problem taken up.

In order, in an individual case, for a hypothesis about disturbances to be meaningful we must first be certain that the devices to be improved are clocks. Thus let four clocks U_1 to U_4 be produced, of which respectively U_1 and U_2 as well as U_3 and U_4 exhibit a constant velocity ratio; nevertheless, both pairs may among themselves show deviations from the constant running ratio. Then hypotheses about disturbing circumstances may be formulated, whereby it depends on any sort of events that are (temporally) observable precisely when the deviations of the pairs of clocks from the constant velocity ratio occur. If no such events are observable, then we must attempt to produce them artificially by varying some accompanying circumstance or other, say, by incorporating still other devices, that need not be clocks, into the comparison. This attempt to recognize other events in addition to the disturbance of the clock pair and thus to find a clue for a hypothesis about the 'cause' of the disturbance need not occur indiscriminately. Since an explanation is sought for differences which show themselves in two given clocks, all further differences (without knowing whether these could be relevant for the running of the clock) must be ascertained, for example the material from which the clocks are manufactured.

Assumed that one has produced two pairs of pendulum clocks U_1, U_2 and U_3, U_4. Thus for each pair let, in the sense of the homogeneity principle, invariability, agreement and equality (by approximation) be given. Now one observes from time to time variations in the running ratio of U_1 and U_3, i.e., there exist *per definitionem* disturbances of both these clocks. Assumed further, that, with the here already presupposed production of geometric forms like straight edges, cog wheels, etc. in connection with the elimination of disturbances with geometric homogeneity principles, one has gained empirical knowledge about the heat expansion of bodies, something that is surely possible without chronometry. One can, then, in the case of disturbed pendulum clocks formulate, say, the hypothesis that the running variations would rest upon a differing, say, materially conditioned temperature sensitivity of the two clocks – the pendulums may consist of different materials.

At this point it is now objected to protophysics that it must give a criterion that is unique in the sense that in a case such as the one imagined it could be decided which of the two clocks is disturbed and which is undisturbed. That, however, is neither possible nor necessary. Each adjustment of one of the clocks to the running of the other and with that the designation of one clock as disturbance-free remains rather a purely arbitrary decision, which shall be banished by protophysics precisely from the foundations of physics, or, on the other hand, it would make use of methodologically unavailable physical knowledge, namely knowledge not available here without clocks. Thus, for example, it would be obvious for a physicist in the imagined case to adjust the running of the clock whose pendulum has the greater heat expansion coefficient to the running of the other clock. It is self-evident that to do this this physicist needs knowledge of the laws of pendulum oscillations. Therefore, in the situation considered here, the disturbance hypothesis can only be: Temperature variations are responsible for variations in the running of both pendulum clocks. This hypothesis may then be considered successful if, e.g., keeping the temperature constant makes the differences in running disappear. Proper names for the measuring instruments under consideration, if they appear at all in this disturbance hypothesis, appear only symmetrically (e.g., the material of U_1 and U_3 is different); asymmetric disturbance hypotheses are untestable (e.g.: U_1 runs slower then U_2 because the heat expansion coefficient of the pendulum of U_1 is greater.) And in fact a physicist would arrange for temperature compensation or a constant maintenance of temperature for both clocks, so that the requirement of a 'unique' choice (selection of one of two clocks as undisturbed or more undisturbed) is not only meaningless in the scope of protophysics, but it also passed by in the practice of the experimental physicist.

This strongly simplified example, which is supposed to describe the 'local' improvement of clocks, may be easily carried over to enforcing the undisturbedness of clocks in the case of relative motion and may naturally be extended to clocks which work according to a principle other than that of the pendulum clock. If, in the case of clocks moving relative to one another, disturbances should occur, even though locally the same clocks show no deviations from constant running velocity, then first e.g., the hypothesis is to be tested whether the disturbance depends on the path travelled. Supposing that the disturbances have been proved to be independent of the path, then there is no objection to the further hypothesis that the motion is the cause of the deviations which occur. Then we must clarify by experiments whether or not, e.g., in particular the acceleration experienced by the clock-pair produces disturbances. Should we be able to demonstrate empirically that hypotheses about velocity or acceleration as a disturbing cause have been established in due order, as e.g., in the case with the acceleration of pendulum clocks in the act of falling, then from this naturally the same consequences must be inferred as with every other disturbance as well: The clocks must be so improved that they are no longer disturbed in the case of motion relative to other clocks.

Since it cannot be required methodologically that a best realization of an 'ideal clock' be produced before measurement of masses and charges and then empirical physics come into play, since, in other words, in the procedure of exhaustion, hypotheses can also be considered which contain terms e.g., of dynamics or of an empirical theory, the problem of a sufficiently good realization of the ideal clock is in fact quickly solved. What the suggested example of the exhaustion procedure ought to show is among other things the following:

1. The propositions exhausted here, i.e., carried out by considering empirically verified hypotheses, in no way exhibit a structure as rich as the theorems of Euclidean geometry. More precisely, they are only propositions of clock-free kinematics as well as those in which the predicates 'invariable', 'agreeing' and 'displaceable' occur.

It would seem to be an explanation if one should wish to add that this simply stems from the one-dimensionality of time. To be sure, the requirements made of the operation of devices are, in a determinable sense, simpler than those which determine the basic form of the plane in geometry. Moreover, not only does geometry restrict itself to *one* basic form directly prescribed by homogeneity principles, but in addition to planarity, at least also requires orthogonality as a basic concept. Talk about 'dimensions' may be introduced only via a thorough discussion of these differences and cannot be elevated to an explanation of them, i.e., a logical presupposition.

2. The procedure of exhaustion does not cease as long as the intention which supports the construction of a physics unique in its propositions is not given up. Thus if someone were to cease the gradual improvement of clocks, specifically there with hypotheses about velocity as a disturbing cause, i.e., were one to wish to abandon the procedure of exhaustion – which would then be equivalent to the assertion that moved clocks are in principle and unavoidably disturbed – then he must state reasons for this. These reasons must be non-circularly intelligible; otherwise this step would have no justification in the methodological construction of an exact theory and would imply propositions that could no longer be defended as true.

The objections resulting from special relativity theory are already irrelevant here for a trivial reason, i.e., because with the chronometry constructed here we are still always concerned with a part of protophysics independent from optical regularities in the widest sense.

In contrast to that, an assertion can be advocated here which argues conversely from the established chronometry to special relativity theory: If the (empirically verifiable with a methodological chronometry) hypothesis of the constant velocity of light in a vacuum and the associated definition of simultaneity at separate places leads to the prediction that clocks moving against one another manifest, solely on the basis of their motion, differences in operation in a time comparison by means of light signals, then from this the conclusion must be drawn that comparison of clocks by means of light signals is an unsatisfactory procedure. In fact no presentation of special relativity theory is known which non-circularly establishes why the attempt to produce undisturbed clocks or at least to define them conceptually was given up. Thus, here, the possibility of a theory unique in the indicated sense has been abandoned without an intelligible reason. (The decisive reason for the physicist is indeed and as in well-known the Lorentz-invariance of the Maxwell equations; but even this reason may not be placed in any methodological context.)

Where the claim to truth in the sense of executability of all individual steps, and, thereby also the claim to the non-circularity of science and here especially of physics, is not to be sacrificed to the questionable gain of including in the theory some phenomena not explained without such a step, then a new definition of simultaneity at different places must replace the unsatisfactory one of special relativity theory. Such a theory, then, in spite of the pretence of having to act under the compulsion of empiricism, may not be taken as empirically founded. The 'extension of simultaneity in empirical space' can be managed non-circularly by means of the transport of undisturbed clocks. This concept of simultaneity is thereby no less subject to empirical control than that of special relativity theory. But, wherever time-measurement sees itself confronted with astronomical distances, this procedure indeed may encounter insurmountable difficulties (at present or always). In that case a theory of light velocity to be gained with the help of undisturbed clocks could be utilized. But, for logical reasons, this would result in no 'time' or 'simultaneity' other than that of undisturbed clocks.

Thus, even with rigorous revision of traditional doctrines, should reasons result somewhere which entail a distinction between the simultaneity of undisturbed clocks and in the simultaneity of light signals, then optical comparisons of time must either be modified or assimilated to protophysical chronometry by means of arithmetical correction of their results.

If, therefore, Hermann Weyl, who knew full well the gaps in the conceptual clarification of the foundations of physics, recommends; that "Philosophical elucidation remains a great task of a type completely different from that which falls to the individual sciences; let the philosopher now see to it; but let one not shackle and hinder the forward progress of the concrete subject domains of applied sciences with the chain-weights of the difficulties which lie in that task (Weyl, 1923, p. 2)", this is not to say that an occasional delay of empirical science which has been free from the chain-weights of foundational efforts, and thus undoubtedly a faster progressing science, is not of advantage. The effectiveness of progress is measured namely not only by speed, but also by the correctness of the path taken. Paradoxically, modern physics from the view of the philosopher striving for an *a priori* protophysics is to be reproached for being oriented precisely too little to 'concrete subject domains'. 'Metaphysical' premisses again enter into the construction of physics via the back-door of grandiose formal systems; these undermine, above all, the claim to empirical confirmation of some presently recognized theories.

4.4. *Outlook and Consequences*

What remains in problems yet to be solved before we may assert that time measurement has been theoretically established is neither a subject of particular differences of opinion nor does it present logically or methodologically difficult problems; consequently, except for a mere cursory mention of some viewpoints, no thorough discussion is required here.

To be sure, a misunderstanding must be prevented: Neither the indication that the ball-groove clock or the pendulum clock is a realization of the 'ideal clock', nor the example illustrative of the exhaustion procedure are to be understood as a recommendation that such clocks be used in practice. The clocks common today fulfill the requirements – to be understood ideatively here – with far higher accuracy than could be attained by mechanical clocks. On the other hand, the example of the ball-groove clock is again not fully superfluous; methodical chronometry could not manage without it or without another, simply and non-circularly realizable, mechanical clock (e.g., water clocks, see p. 168). Since 'methodical' shall imply as much as 'reanactable' (*nachvollziehbar*), and since the postulate of non-circularity permits no knowledge gained earlier to be incorporated into the execution – it is precisely a question of distinguishing knowledge from pseudo-knowledge, each according to whether it may be established methodologically or not –, we must provide for the construction of primitive clocks in an execution of chronometry. Chronometry then furnishes presuppositions for a protophysical theory of mass measurement in which laws of oscillation are derivable for the case of a homogenous force-field. If, empirically, on a small scale the earth's field of gravity is recognized as a sufficiently good realization of such a field, then nothing stands in the way of using pendulum clocks. Analogously may we understand the improvement of clocks achieved in history, from the tuning-fork clock and quartz clock up to the cesium clock.

First, chronometry has heretofore produced only a designation of a *form of motion*. The second step is the introduction of a metric which may be taken over unchanged from geometry. With this no more difficulties occur than with the choice of a unit, for which, in addition to being a magnitude satisfying practical requirements, the most accurate and simplest possible reproducibility is decisive. The stipulation of units of magnitude does not belong to protophysics. It is empirical knowledge which chiefly enters into their choice. What then may be considered as having been methodologically established is a *kinematics* in which (in contrast to the usual presentations in textbooks) no pure temporal propositions occur.

Due to its brevity and simplicity, the generally familiar manner of speaking of time is of some advantage. Moreover it has its place in everyday life, where clocks are used unreflectedly. (There, to be sure, its uncritical employment is adequately justified by its practical usefulness.) We shall therefore indicate how a manner of speaking of time can be obtained from the methodological chronometry constructed up to now.

Pure temporal propositions are, in short, to be understood as propositions about motions of clock-hands, where what clock we are directly speaking about and what geometric peculiarities it has are taken to be irrelevant. Talk of time is, in other words, gained by abstracting from chronometry, where here indeed the very name 'chronometry' is misleading. More accurately, it is a kinematics from which propositions 'about time' are abstracted. A look at the usual propositions in ordinary language allows us to recognize a rough division into those propositions which concern a 'durational aspect' of time. Events are compared in respect to their duration and in respect to their succession in an actually occurring or imagined experiential context. Durations, in particular, are placed into a numerical relation. (A reference to some sort of standard event for determining a duration unit or for fixing a null-point, as in calendar reckoning, is of no importance here.)

Reconstruction of both these propositional groups from a methodological theory of motion starts out from the concept of the ideal clock. In particular, let us here emphasize the condition that the ideal clock 'always' runs; this is equivalent to the fact that propositions of a temporal sort are possible for every event, regardless of whether it is real or only imagined.

The relation of two durations is then given by a rational number which may be interpreted as the path-length ratio of two events of the hands of the ideal clock which are simultaneous to these durations. This type of 'correspondence' of numbers to pairs of events starts then from the procedure of first designating as equally long simultaneous events with a simultaneous beginning and simultaneous end; this procedure itself cannot be chronometrically analyzed further. Thus talk about durations of simultaneous events may be comprehended as an abstraction from talk about concomitant events. The procedure of comparing two arbitrary, non-similtaneous events in respect to their duration encompasses three steps: For each of both events a concomitant event is indicated on a clock, and these are then compared in regard to their path-lengths.

In a historically conditioned and for the most part ununderstood manner, one also, in addition to durations, speaks about points of time. This tradition, which begins with Zeno and Aristotle, occurs in two variants; one of these goes back to Augustine, according to whom the point of time separates the future from the past – the ordinary-language use of 'point of time' or 'moment' may have arisen from this conception which refers to an 'experienced time'. The other variant has a more natural-scientific character; according to it a point of time is the limiting case of increasingly smaller divisions of time. Without becoming aware that the analogy to the geometric 'point' is unsound as long as geometry is developed under a rejection of terminological definitions for the basic concepts, natural scientists talk of points of time in connection with the mathematical aids they employ. Nevertheless, even in physics, the phrase 'point of time' (and its synonyms) could be abandoned without having to trade for it a peculiar artificiality in presentation. If, all the same, one wishes to take the tradition into account, then one must systematically reconstruct the talk of points of time.

Points of time, in contrast to durations, cannot be abstracted from occurrences. The words 'occur' and 'point of time' are generally used in just such a way that nothing occurs in a point of time, i.e., every occurrence requires a finite duration. To avoid disturbing associations with geometry, let us therefore first propose the word 'instant' instead of 'point of time' (*Zeitpunkt*).

In the same manner as with 'duration', one uses 'instant' to circumlocute the adjective 'concomitant', when instead of 'the positions s_1 and s_2 are concomitant positions' one says 'positions s_1 and s_2 are occupied at the same instant'. Where concomitant positions are not spoken of, instants are abstractions from positions. By restricting oneself to propositions about the ordering of positions to one another, one obtains propositions about instants. As 'ordering predicates' the following can occur here e.g.: 'between' (position s_1 is assumed between s_2 and s_3) and 'situated as' (s_1 is situated to s_2 as s_3 is to s_4). In everyday speech, where the immediate experience (*Miterlebnis*) of events produces an orientation of time, the predicates 'earlier' and 'later' supervene. Finally, then, rules for durations and instants analogous to those for events and positions may be set in force, according to which instants are the bounds of durations, etc.

With this the talk of durations and instants, obtained by abstraction from propositions about events and positions, is intelligible. We abstract from the geometrical peculiarities of paths and places as well as from the choice of the bodies examined, which in particular implies that which real clock is chosen is disregarded.

A reference to two problem areas which no longer belong within the methodological bounds of an *a priori* theory of time-measurement constitutes the conclusion of chronometry. Even though this is the case, they are occasionally discussed in connection with chronometry; thus a demarcation from them appears necessary.

In clarifying the preliminary understanding which was brought into the construction of a chronometry, we pleaded for not including the predicates 'earlier' and 'later' among the terms of the theory. Now after kinematics and, beyond that, pure temporal propositions have been reconstructed without this word pair, it can be taken as demonstrated that it is not necessary for a methodological chronometry that we supply 'time' with an orientation.

If one treats bodies and events analogously in the indicated manner, i.e., if with repetition of 'the same event' one speaks of how one can repeatedly encounter the same body, then the temporal succession of two events is reversible in the case of repetition. In physics one can agree, in a second step, (which however should not cause us to forget the first step!) to locate *bodies* solely by means of place coordinates, and *events* however by means of place and time coordinates, thus obtaining an asymmetry in the bargain. This then results from considerations of expediency alone and need give no cause for additional reflections 'on the nature of time'.

Consequently, a chronometry which does not define a 'direction of time' – something which would always be possible only as an invariant form of talk about events – makes no prejudgment relative to the question of whether definite events are 'reversible'; this question occasionally surfaces in empirical physics and is important in its consequences. After the discussion of the conceptual foundations of chronometry it can be further asserted: The question of the reversibility of events, which makes use of a qualitative definition of 'states' and, to that extent, in most cases already presupposes chronometry, cannot itself be relevant for the foundation of chronometry. To terminologically indicate this independence of methodological chronometry from the empirical question of the 'reversibility' of events, let us call the present chronometry *reversible*. The 'reversibility of time' or also the 'direction of time' may be (in an unsuitable terminological phrasing) a genuine problem of empirical physics; in protophysics it may remain unnoticed.

The second problem, which is not to be treated in chronometry (to be sure, though, in protophysics!), consists in the question of transformation equations which comprehend the transition from one reference system to another. In current physics this problem is seen as a kinematical problem into whose solution the theorem of the independence of the speed of light from the reference system, which is derived from an interpretation of the Michelson experiment, enters. In fact the stated experiment is not 'dynamic'.

However, one must distinguish a reference system determined via optical experiments from a kinematical one. The optic-independent transformation properties of a *kinematic reference system*, i.e., of a geometrical coordinate system marked on a rigid body by four points not lying in a plane and the clock connected with this system, depend solely on the ideatively required constancies of the body and of the clock. Where, then, indeed 'dynamical reference systems' are defined, as say by the condition that in a dynamical reference system a body uninfluenced by forces moves uniformly and in a straight line, the bounds of transformation properties controlled by kinematics have been overstepped.

The present chronometry can, stated briefly, provide no sort of decision as to whether the 'laws which govern reality' are Galileo- or Lorentz-invariant. The 'laws which govern reality' are *qua* linguistic actions – which in any case they also are – fully dependent on the way man arrives at them. Methodological chronometry provides a first and, in view of the problem, short step when it formulates requirements which enable the realization of Galilei-invariant reference systems. The justification for starting precisely with *rigid* bodies and *undisturbed* clocks is the unique reproducibility of constant measuring devices, a problem that is not taken account of in relativity theory and yet decides the intersubjective verifiability of empirical sentences.

CHAPTER IV

ON A HISTORY OF CHRONOMETRY

1. PRELIMINARY REMARKS. TERMINOLOGICAL DISTINCTION OF PRACTICAL AND THEORETICAL CHRONOMETRY

The theory of time-measurement developed in Chapter III not only represents a clarification of the conceptual foundations of physics. It also offers, by means of standardized terminology, a proper instrumentarium for a historical investigation of chronometry. We may have been thereby justified in postponing such considerations until this chapter.

For a large part of their history two developments proceed nearly uninfluenced by one another and only in the 17th century, with the rise of classical physics, do they enter into a fruitful alliance with one another. They are practical time-measurement and theories of time.

A preliminary demarcation of both these branches, which are hereafter called 'chronology' and 'time theories', suggests that 'chronology' denotes the unreflective practice of arriving at temporal assertions with the help of observations of natural or artificial processes; while 'time theories' are systems of propositions which – as is frequently said in the tradition – set forth assertions about "the nature (*Wesen*) of time". The development of chronology moved forward out of more or less practical needs; its patrons were statesmen, priests, architects and engineers. The tradition of time theories, however, has its home in the philosophical schools.

Let us indicate here at the outset the limitations under which the philosophical efforts are considered: the history of a 'philosophy of time' will not be reported. Rather, on the basis of the understandings explained in the systematic part of this book, the history of a concept of time will be presented as that concept is to be found in classical physics and thus in the natural sciences in general.

Since within the scope of this work we are only striving to contribute to an understanding of the modern use of clocks and to answer the question of what end the watchmakers may have 'always' pursued in building clocks, the consideration of the history of chronology also remains restricted to a narrow domain.

Just as unpretentious prescientific speech methodologically precedes the high stylisation productive of a scientific language, so the practice of chronology historically precedes theoretical attempts to elevate time to the 'height of the conceptual'. This order of succession is analogous to the genesis of Euclidean geometry, which, as a theorization of land-surveying, is to be placed later in time than the practice of the surveyor itself. It was in the step from the practice to the theory of spatial measurement that that discovery of the possibility of science (Cf. Mittelstraß, 1965) occurred, a discovery which – so much may already be asserted now – has never properly come to fruition for chronometry. Thus a history of chronometry appropriately follows the historical sequence and begins with chronology.

2. THE DEVELOPMENT OF CHRONOLOGY

Chronology may be divided into 'time reckoning' and 'time-measurement'. In conformity with common usage, 'measurement' always implies the use of measuring instruments. Thus wherever men have recourse to manfactured instruments in order to obtain statements pertaining to time, we shall speak of time-measurement; statements pertaining to time without relation to instruments belong to time reckoning, as, for example, the calculation of calendars.

The history of time-measurement suggests a further distinction: Employment of instruments for observing natural processes, i.e., processes such as those which have not been set in motion by man (the revolution of the sun for example) belongs to 'natural time-measurement'; while 'artificial time-measurement' encompasses the production of the observed processes).

Since the presuppositions of artificial time-measurement have been clarified in the systematic part of this work, the historical consideration of chronology will likewise be limited to artificial time-measurement.

A look at time reckoning and natural time-measurement is indeed worthwhile where it is a question of what gave rise to chronology and to time-measurement in particular. Answering this question is relevant for judging individual phases in the development of time-measurement.

Proper names for seasons, months and days (annually recurring holy days, and only later week days), names from a time out of which stems the oldest evidence of a beginning of time reckoning, indicate that the beginnings of a farmer's calendar progressed with the development of agriculture.

With the division of the natural year in Egypt, with its three seasons 'flood', 'sowing' (winter), 'harvest' (summer), into twelve months, some monthly names were selected in accordance with seasonal or agricultural practices. (Cf. *Lexikon der alten Welt*, Zürich and Stuttgart 1965.) According to the literature, administrative tasks (tax dates) and the fixation of cultic feasts accounted for the introduction of a calendar year based on the movement of Sothis (Sirius) (presumably in 2780 B.C.) (Cf. P. Lorenzen, 1960, 21). In Egypt the need to fix the date of the Nile flood as accurately as possible in advance also supervenes.

In the Babylonian reckoning of time (the calendar of Nippur) a six-part division of the month into five-day business weeks may be surmised (Cf. *Lexikon, op. cit.* p. 1465); in the Greek calendar an intercalary cycle (*Octaeteris*) was introduced by the priests of Delphi in order to be able "to bring the same yearly fruits to the gods at the same times" (Cf. *Lexikon, op. cit.*, p. 1466); even the beginnings of time-measurement seem to have been performed primarily for the fulfillment of cultic instructions – an analogue to the employment of the first mechanical clocks in the medieval cloisters. Finally, strict organization of the army required usable clocks for dividing up the night watches.

This orientation of chronology either toward the organizational requirements of daily life or toward cultic instructions may be traced from the first beginnings of chronology right into the Middle Ages. Moreover, the use of clocks for scientific purposes, i.e., for astronomy, already appears in ancient Egypt. A. Lübke writes (Lübke, 1958, p. 73) that "the Egyptians accomplished astounding things in the use of water clocks for astronomical measurements", say, the determination of the perihelion and aphelion respectively.

A more effective use of clocks for astronomical measurements could, succeed it is true, only after the invention and definite perfection of mechanical-escapement clocks (presumably not before the 15th century).

Clocks appear to have been first used especially for experimental investigations by Galileo (see time-measurement by Galileo, p. 179). He not only laid the beginnings of an exact empirical science by uniting the tradition of the workshops with that of the universities;[21] he also surpassed his predecessors precisely in the respect that he introduced the time parameter as representation of an experimentally measurable time into physics.

Where now is the beginning of an artificial measurement of time to be recognized? According to the terminological prescription, sun dials – initially doubtlessly used as 'calendars', i.e., as instruments for indicating the position of the sun in the signs of the zodiac – served for a natural measurement of time. To be sure, due to their easy transportability, they were also in use after the invention of the water clock (upto the 18th century). The hourly division of the solar day (in Egypt demonstrable from 1400 B.C.) may also have been initiated by means of a twelve-part sun dial scale. However, since there was no theory of the gnomonics* of sun dials until around 350 B.C. (Eudoxos of Knidos), a measurement of time which could relate arbitrary durations to one another was theoretically not possible. Presumably even sun dials were empirically calibrated with water clocks. (No unanimous opinion prevails among the experts as to whether the production of an '*arachne*' presupposes knowledge of the principles of stereographic projection. In any case, Eudoxos appears to be the inventor of a *ὡροσκοπεῖον*, a sidereal clock which consisted of a pierced disk and indicated "the zodiac and the parallel circles, as far as they were necessary for determining the most important stars" (Lasserre (ed.), 1966, p. 159). This device, which earlier recalled a cob-web, was later simplified by omitting the parallel circles and in this form resembled a spider, from which the name '*arachne*' explains itself.)

The use of water clocks constitutes the beginning of artificial time-measurement and thereby the beginning of a rigorous measurement of time in general.

3. SHORT HISTORY OF THE WATER CLOCK

It is difficult to date when artificial processes were first turned to for measuring time. Nevertheless it is generally agreed that the oldest procedure was one which foresaw letting water run out of a container through a small hold. In a grave inscription from around 1580 B.C. (Cf. Taton, 1963, p. 41), Amenemhet gives instructions for the construction of a water clock; the oldest preserved specimen comes from around 1350 B.C.[22]. However the simple form of the outflow clock is presumably much older. It already appears to have been in use among the Chaldeans.

The earliest form of the water clock merely allowed determination of a definite duration through the emptying of a container, without enabling one to read smaller divisions of this duration on a scale. (Cf. Sarton, 1959, II, p. 344). In Egypt such instruments served to limit the time of speeches in court; this was done by pouring more or less water into the drainage receptacle, depending on the seriousness of the case being tried, at the beginning of the trial.

The Greek clepshydra (water siphon) worked according to the same principle. (Clepshydra – water thief – at first served certainly to take water out

* Here '*Liniennetze*' is translated as 'gnomonics' since this is the term in the literature for the science of constructing dials so that they can compensate for differences in latitude; it involves both the positioning of the style and the tracing of the 'hour lines' in such a way as to get accurate solar time – Tr.

of cisterns. Their manner of functioning is comparable to that of modern pipettes. Thus later the name clepshydra became common for water clocks in general, even for those with scales.) It was also traditional in Attic judicial practice to limit speech times by means of a clepshydra. Scaleless outflow clocks already found early application in dividing up the night watches. Among these early forms of water clocks the fact that the velocity of the water flowing out of the container varied with the height of the water level was not relevant. However, the construction of water clocks with scales (the oldest specimen, mentioned above, has twelve different scales – corresponding to the number of months –; each scale yields a division of the day into twelve equal hours with respect to the length of the solar day (Cf. Taton, 1963, p. 41) required a solution to the problem of producing some movement or other with a constant velocity – uniform sinking of the water level or constant outflow velocity for example –, such a movement yielding a uniform rise in the water level of a cylindrical collection receptacle.

In rough detail this was accomplished in the case of the preserved specimen by conically widening the shape of the container near the top instead of keeping it cylindrical; this should have compensated for the increased outflow velocity in the case of high water level.[23]

As far as is known, the problem of finding the correct container shape which would have enabled a constant sinking of the water level (surface parabola of 4th order, $x = 4y^4$) was not solved in antiquity. Nevertheless the construction of accurately running water clocks was accomplished when Ktesibios of Alexandria (around 300 B.C.) devised and also carried out several constructions in which the utilization of at least two receptacles was common.[24]

In the clocks of Ktesibios constant outflow velocity was attained by means of a valve controlled by a float or through the preswitching of an overflow container. Then, with the help of a float, the uniform rise in the water level of a cylindrical collecting receptacle was transmitted to various indicator mechanisms.

After Vitruv transmitted detailed instructions of Ktesibios, which were to be heeded in order to achieve a constant outflow velocity (like utilization of clean water, or of corrosion resistant material for the drainage hole) the main interest of ancient clock builders was focussed on indicator devices, of which a large number of playful forms have been handed down. Besides acoustic signals from pipes or balls, which struck the hour by falling into bronze basins, clocks were coupled with a sort of mechanical astrolab which gave information about the seasonally visible starry sky or gave the position of the sun. The chief difficulty in the construction of water clocks lay in the varying length of the day which was to be divided into twelve hours of equal length. (The division of the full day into 24 equal hours, already known among the Babylonians, could not make general headway for a long time. As late as + 1350 both ways of reckoning hours were still in use (Cf. Zinner, 1954, p. 11).) Instead of variable scales – already Ktesibios had constructed an automatic shift for the cylinder which bore the different scales – attempts were made to vary the outflow velocity; this was accomplished by varying water levels in the reservoir, by varying the position of the drainage openings or by gradually opening and closing the drainage openings.

Since water clocks were in use as late as the 18th century, they appear to have well fulfilled the demands placed on them. (Cf. Crombie, 1952, p. 185.) In the beginning of the Middle Ages, they belonged, as it were, to the basic equipment of a cloister, since the rules of the order exactly prescribed the prayer times. It was here, presumably, that the deficiencies of the water clocks were felt most clearly – they required most laborious maintenance and they froze up in winter.

4. SHORT HISTORY OF MECHANICAL ESCAPEMENT CLOCKS

The accuracy of water clocks, in which the time was given through a variable level of liquid – either directly or in connection with an indicator device – could not be increased at will. The accurate time-measurement needed by physicists and astronomers, and later in technology, was bound up with the invention and development of mechanical clocks.

If here now the history of the mechanical clock is traced, in view of the role which it played in the history of exact empirical science and which it still plays in spite of quartz and cesium clocks, then this need not give rise to the impression that, historically, it immediately superseded the water clock. Though mechanical clocks were furnished with minute hands as early as the 15th century, their accuracy did not become satisfactory until the days of Huygens, Hooke, let alone Harrison.[25] Thus, besides many other forms of chronometers – mercury clocks for instance –, water clocks themselves were still long used for astronomical observations. Kepler mentions a 'cylinder clock' (*Walgeuhr*)[26] constructed by Salomon de Caus; and as late as the 19th century, Clocks were built which combined principles of water clocks with wheels or even pendulum mechanisms. (Cf. Lübke 1958, pp. 78–84.)

Before our investigation of the beginnings of the mechanical clock and its first usable examples, we will offer some remarks about what, besides enjoyment of playful mechanisms, may have been decisive grounds for the development of time-measurement. (Similarly to the first water clocks, the first mechanical clocks also had many indicator devices. Frequently they were conceived less for indicating the time than for operating a type of planetarium. Show clocks, in which doors opened, puppets or animal figures appeared, etc., enjoyed great popularity, as well as clocks coupled with chimes or with musical instruments of all sorts.)

With the development of new forms of ships (utilization of lateen sails instead of the old square rigging, several masts, steering rudder on the stern post, which was itself an extension of the keel) travel on the open sea increasingly replaced coastal navigation. The first charts used after the invention of the magnetic compass (around 1200) showed only coastal regions with landmarks which could be sighted from the sea; they contained, however, no indications of distances. Navigation on the open sea was dependent on determining the ship's position. Latitude could be determined by taking readings from the stars; however, for determining longitude a clock would have been needed. Comparing the temporal shift of the solar day with the time shown by a clock brought along on the voyage (perhaps the time of the home port or of a world-wide reference point, such as Greenwich served after 1884) permitted calculation of longitude. Even if this problem may hardly be considered solved before 1750 (see note 25), the production of an accurately running ship's clock was at any rate already a goal of clock builders.

In addition, it was not only a small circle of astronomers who needed more accurate and thus more workable clocks. As early as 1400 it became the general custom to pin down as accurately as possible the birthtime of a child, since this had to be known for casting a horoscope.

The first steps on the path to weight-driven mechanical clocks are closely related to the technology of that time, especially to manufacturing technology and the development of special tools, above all the development of the simple form of the lathe. Since wood was shown to be little suited for clockworks – though wooden works were in use for a long time – even pocket watches were made from wood[27] – the development of mechanical clockworks began only with the utilization of metals; initially wrought iron was the most used.

Here clock building took on a role which is not to be underestimated for the construction of tools and the technology of metal manufacture in general. Besides fire-arms, clocks were the only metal devices in the 14th century whose production demanded some precision. If the first mechanical clocks were still products of the blacksmith,[28] the realisation that clocks work all the more accurately the less the active forces are, soon led to a diminution of the works. The development of the lathe and other tools needed for the production of shafts, cog-wheels, and such combined with this.

The history of the mechanical clock is essentially the history of escapement mechanisms, which were to brake a falling weight in such a way that the fall took place sufficiently slowly and uniformly. For this, in the literature, the French master builder Villard de Honnecourt is the first mentioned. (Since 1955 some scholars are indeed of the opinion that a first example of the use of escapement mechanisms can be seen in the colossal clock of the Mandarin Su Sung, a water clock containing an escapement mechanism (1088). In fact the escapement mechanism of this clock resembles a form also appearing in European cog-wheel clocks (see J. Needham, Wang Ling, D. J. de Solla Price, 1960.) However, according to *Die Zeit*, p. 79, there is no proof that Europe acquired knowledge of this clock or of its predecessors which have been dated up to 300 years earlier (around 800).) Villard de Honnecourt designed an 'angel clock' (around 1240); the clock was a weight-driven apparatus furnished with an escapement mechanism and was to have rotated an angel in such a way that it always remained turned toward the sun.

Zinner describes this clock (Zinner 1954) and explains its functioning, in agreement with other standard works,[29] as follows: "The escapement takes place through the weight of the wheel, which can be selected in such a way that the revolution of the wheel through 1/4 of its circumference, i.e., from one spoke to the other, corresponds to the day's length." Though, as far as Villard's clock is concerned, one remains solely dependent on attempts to explain his drawing, this is not very convincing.

The revolution of the wheel W and subsequent rotation of the shaft A, which was to have borne the angel, is accelerated uniformly. It does not seem to have been Villard's intention perhaps to hinder this movement by friction in such a way that it would proceed with constant velocity. Otherwise neither the large wheel nor the leading of the rope through the spokes of the wheel would have an explanation. Granted that Villard could not have known the law of gravity, one might nevertheless credit him with a qualitative knowledge of the movement of a simple mechanism. Moreover, it is unpleasant to impute to a master-builder, a practical man (who was thoroughly successful in practice, as some of his constructions, for example, a sawmill driven by a mule, prove (Cf. White, 1965)), the conception of the possibility that such a slowly running mechanism (losing 1/4 revolution per day!) might function.

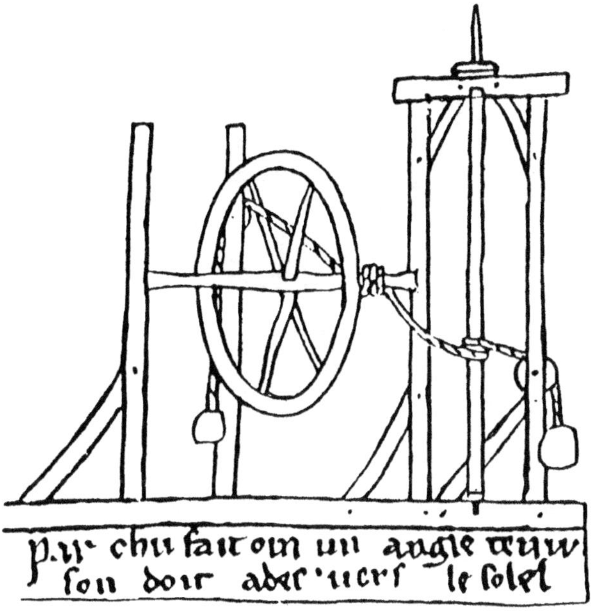

Fig. 20

By far the most weighty objection to Zinner's explanation is that he does not convincingly treat the question of why the rope was led in between two spokes.

In the following, then, a new explanation of the functioning of the angel clock shall be proposed; according to this explanation the wheel W oscillates back and forth and thus can already be viewed as an escapement mechanism regulating the falling motion of a weight by means of its periodic operations.

In a simplified sketch, the mechanism (the shaft bearing the angel has been omitted) appears as follows:

The rope R runs from weight G *in front* of the wheel up to the shaft S, winds around this and then runs between spokes 1 and 2, over a pulley, to weight g. In the position indicated the wheel is accelerated counterclockwise by a force

$$F = (G - g)r_S. \quad (r_S \text{ is the radius of the shaft } S.)$$

Finally spoke 1 hits against the rope. When the swing of the wheel is absorbed, end position I will have followed:

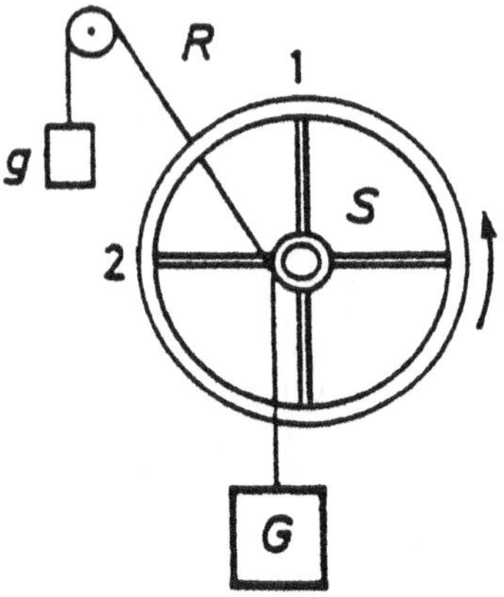

Fig. 21

In this position the backward propelling force is

$$F = r_{w_1} g - r_S G.$$

The wheel will then be accelerated clockwise and finally reaches end position II.

Now $r_{w_2} g$ and $r_S G$ work in the same direction; the wheel is thus accelerated counterclockwise again, initially with the force

$$F = r_{w_2} g + r_S G,$$

until the rope no longer touches any spokes; from there on it is accelerated with the force

$$F = (G - g) r_S$$

wherewith the oscillation begins anew.

In end position I, the rope, by way of example, lies along 3/4 of the shaft circumference; in end position II, on the other hand, it lies along 1 1/4 of the circumference. With a suitable adjustment of the individual components to one another it is possible that the static friction of the rope on the axle will not suffice to maintain $G - g$. The rope will slip somewhat[30] until the wheel, now only accelerated by the force $F = r_{w_1} g$, has again revolved so far clockwise that the rope catches hold of the shaft. Theoretically the kinetic energy recovered by the slippage of the falling weight G would suffice to offset the frictional losses of the oscillating wheel.

Although this explanation is physically correct, one immediately sees that

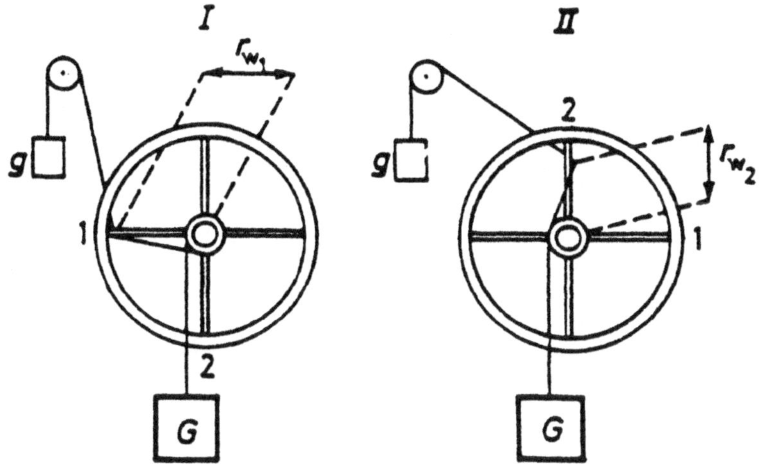

Fig. 22

Villard's mechanism will not function in practice as a usable clock: Even with modern means frictional forces can not be controlled as exactly as the functioning of this clock would require. Above all, the requirement that the rope must slip through at exactly the same length with every period (this length yields the effective rotation of the angel, which is coupled with the rope in such a way that the rope will not slip) is unfulfillable in practice without the employment of auxiliary devices.

It may nevertheless be assumed that Villard had already thought of an escapement mechanism functioning in principle by means of an oscillating body.

A mistake, which, to be sure, does not concern the escapement mechanism, but which indeed does concern the purpose of the whole apparatus, appears to have struck neither Villard himself nor the historians who have described his clock. It is a declared goal of the device to so rotate an angel that it was constantly pointed toward the sun. Then, however, (if one considers it) the shaft A must turn clockwise. If it can be concluded from the clear size difference of both weights that the one drawn larger is also the heavier one, then, in his drawing, Villard wound the rope in the wrong direction round the shaft A. However one may explain the functioning of the clock, according to the drawing the angel will turn from west to east, contrary to the movement of the sun in the northern hemisphere.

In 1271, in a treatise concerned with the division of the day into equal hours, Robertus Angelicus describes what conditions a clock driven by a lead weight must satisfy in order to indicate the hours correctly.[31]

However, Zinner (1954, p. 7) points out, justly so, that this treatise is not concerned with an escapement; thus this passage is unfruitful, (at least) for the history of clock building.

At this point one readily cites Dante's *Divine Comedy* (between 1316 and 1321).

> "And even as wheels in clock escapement ply
> In such a fashion geared that motionless
> Appears the first one, and the last to fly."
> (*The Divine Comedy of Dante Alighieri*, translated by Melville Best Anderson, Heritage Press, New York 1944.)

However, all this is only an admission that one knows neither the inventor of the mechanical escapement clock or of the verge escapement nor the date of these inventions. At this point it may suffice to remark that the first mechanical clocks with verge escapement[32] were presumably built around 1300 and that the watchmaker's art already had a high state in the 14th century; this can be observed from some still partially preserved clocks of the second half of the 14th century, clocks with automata, chimes, alarm contrivances, etc.

The spindle clocks, initially possessing only two or three cog wheels, had the disadvantage that they had to be wound up often. Their accuracy left much to be desired. ("Accordingly, clocks (those with verge escapement) were deemed good when they gained or lost no more than 15 minutes a day." (From *Die Zeit*, p. 80.))

Thus various forms of clocks remained in use as late as the middle of the 17th century. To be sure, works with verge escapement could be improved substantially; however exact measurement of time was first to become possible with the regulation of clocks by a pendulum. About half a century before Galilei and Huygens introduced time-measurement by means of pendulum clocks, two escapements were conceived which resulted in significant progress in running accuracy as compared to that of spindle clocks. Besides other improvements, Jost Burgi invented the 'cross-beat' which suppressed running irregularities by means of a symmetrical arrangement of two arms striking against each other; and between 1590 and 1600, clocks with ball-drive escapement came into being. Because Burgis's inventions are accurately described in the technical literature, (Cf. E. v. Basserman-Jordan, 1961) they will not be gone into in more detail here. On the other hand, ball-drive clocks are of special interest here since – older than pendulum clocks – they are realizations of the ideal clock described in the systematic part of this book, as a device to attain constant velocity.

"Christoph Margraf, watchmaker to the court in Prague, proposed the ball drive on an inclined plane as a time standard to Kaiser Rudolf II" (Basserman-Jordan, 1961, p. 368) and thereupon received a patent. It is probable that this is the earliest evidence of the use of a ball-drive escapement; yet since various uses of balls in watchmaking appear around the same time, fixing the date here encounters special difficulties. For one thing clocks were constructed in which balls kept the *driving force* constant; for the other, the ball drive was occasionally only decorative, having no influence on the working of the clock. Not to be left out were attempts to build, or at least simulate, a perpetual motion machine via a ball drive. One of the oldest preserved examples of a ball drive clock is the box-shaped table clock signed "Christoph Margraf can privilegio Rom. Sac. Maj. 1596, Prag": (Bassermann-Jordan, 1961, p. 132) the 'Babylonian tower' of Hans Schlottheim represents an especially splendid clock. In it the balls run outside around the tower on a spiral track (around 1599) (Cf. Maurice, 1968, p. 40).

Because of its similarity to the form of a methodologically primary clock, theoretically postulated in Chapter III, let us mention here in particular the ball-drive clock of the Ulm watchmaker Johannes Sayller (see picture). A. Weyermann describes this clock as follows (Weyermann, 1829, p. 449): "This clock has a free balance wheel escapement and is set in motion by means of two balls. The balls alternately roll down a slanted glassy surface along six transversal lines formed by taut metal strings; at the ends of the strings they disappear into five semi-circles. Each ball falls onto a balance 180 times in one hour and each is lifted up again by means of the balance just as often. The clock indicates minutes and hours, the day of the month, the phase of the moon, strikes quarter hours and hours; if one of these balls is removed the works suddenly come to a standstill." We add that the balls run with constant speed after having been accelerated on top of a starting track about 3 cm long.

It is true that the limits of running accuracy are quickly reached with ball-drive clocks. The release mechanism for the start of the balls offers special difficulties and a minor cover of dust on the ball race leads to perceptible irregularities in the clock's movement. Only pendulum clocks could fulfill the demands of that time for accurate time-measurement.

As for the history of the subject 'invention of the pendulum clock', the historians are primarily occupied with deciding the dispute as to whether Galileo or Huygens may claim the renown of being the inventor of this instrument so important for the science of the following century.[33]

Ball-drive clock of Johannes Sayller, Ulm 1626, Württembergisches Landesmuseum Stuttgart.

In a synopsis which omits the unimportant details as well as avoids conjectures about the motives of the participants in the not always completely transparent events, the following may be asserted as certain: Galileo recognized very early, probably not later than 1581, that the pendulum was suited for measuring short times. Up to his death he appears to have believed that the oscillation period of a pendulum was independent of the amplitude.

The pendulum, which was hand driven and whose oscillations had to be counted by an observer, came into use among astronomers very shortly thereafter. Knowledge of this use reached Holland.

On the 15th of August, 1636, Galileo wrote to the General of the United Provinces of the Netherlands and suggested a new method for determining distances at sea with the utilization of a chronometer. To be sure, Galileo's letter contains no information about the type of chronometer ("*orologio per numerar l'hore e sue minuzie*" and "*misurator del tempo*", Galileo. *Ed. Naz., XVI*, p. 466 and 467) in question; nevertheless the remark that, if one let four or six of these chronometers run simultaneously, they would not differ by one second, even a period of months, is startling.[34]

It could indeed be a question here of a consequence from Galileo's notion of the isochronism of pendulum oscillations. However, with these chronometers – which were to run over a period of months –, it is improbable that Galileo intended to suggest simple pendulums that must be driven by hand and whose oscillations have to be counted by an observer. Rather, this passage ought to count as evidence that Galileo already conceived a pendulum clock in 1636 – that is, unless he even constructed one or commisioned someone to construct one, since he expressly says "... *io ho tal misurator del tempo* ..." [I have such a chronometer ...] . It is provable that Galileo, together with his son Vincenzio, designed an escapement mechanism controlled by a pendulum shortly before his death when he was already blinded. According to Bedini, this construction, which shows no clock face, served only to count pulses (*pulsilogium*) and to measure short durations in astronomy.

In 1649 Vincenzio either built or had built such a pendulum clock (presumably spring driven). Then, based on this model. Viviani, the biographer of Galileo Galilei, drew sketches which are still preserved today.

Apparently both Viviani and Prince Leopold de Medici,[35] who later defended Galileo's priority claim, did not recognize the significance of Galileo's invention and therefore did not make it public. Later both these admirers of Galilei would make up for this mistake, and not always by correct means, as the response found by Huygens' clocks taught them better.

Huygens, who, by his own admission, first thought of the new method for measuring time toward the end of the year 1656 (Robertson, 1931, p. 86), applied for a patent for the pendulum clock in June of 1657. In 1658 his 'horologium' appeared along with a description of this clock which, however, still did not exhibit the famous cycloidal pendulum.

Huygens first discovered the isochronism of the cycloid after 1659, whereupon he drew a pendulum suspension between cycloidal cheeks in his *Horologium Oscillatorium* in 1673.

Robertson found a clock by Samuel de Coster, Huygens's watchmaker, from the year 1657; this clock as well already shows a bifilar pendulum suspension between two cheeks (not cycloidal to be sure) (Cf. Robertson, 1931, p. 78). Touching now on the question of the priority of Huygens or Gailiei, we see that the circumstance that Huygens's publications date later than Galileo's death (January, 1642) is still no argument for the thesis that Huygens committed plagiarism. In such an argument it appears as though a causal connection has been constructed from a temporal sequence. Admittedly the dispute, in which Huygens himself passionately participated, suggests assuming this causal connection; however, there exists nevertheless no reason to doubt Huygens's protestations that he made his invention independently of Galileo.

Belief in this independence becomes easier the more one disregards the pronouncements of partisans engaged on one side or the other and inspects the constructions themselves. Here a technical argument can provide clarity much more than a painstaking study of the sources and documents concerned with the priority dispute: the escapement mechanism of Galileo's clock is completely different from that of Huygens' clock.

In principle Galileo's escapement mechanism works as follows (see Figure 23): the lever *H*, which is rigidly joined with the pendulum *P* revolving in *A*, gears into the pins *s* of a cog wheel *R*. The pendulum is driven toward the right. The pallet *K* prevents the wheel *R* from turning further when *H* has left the reach of *s*. With the return of the pendulum towards the left end position, the pallet *K* is raised by the pivot *Z* and the pendulum is given a new push via the pins *s* and the lever *H*.

However, Huygens' escapement is exactly that of the spindle clock (see Figure 24): The shaft of the pendulum, furnished with two pallets, stands in fromt of a crown wheel and is oscillated back and forth by the teeth of the wheel.

While the pendulum in Galileo's clock is accelerated only along one part of its path (in Figure 23 only in the region of the left reversal point) and can then swing free, Huygens' pendulum is always in contact with the crown wheel. (Whenever S_1 is no longer touched by the crown wheel, S_2 must already have moved within reach of the teeth of the crown wheel; otherwise the crown wheel could revolve freely.)

The only feature common to both models is the utilization of a pendulum and thus of a structure capable of oscillating with its own frequency

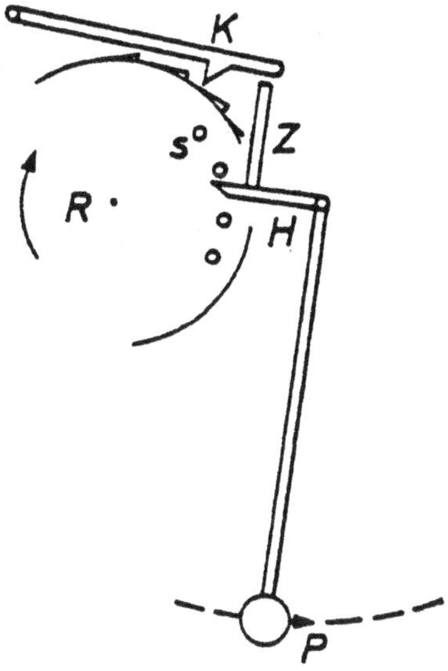

Fig. 23

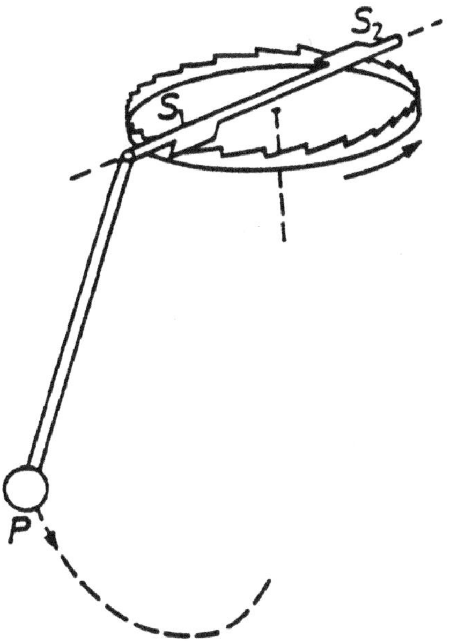

Fig. 24

independently of the driving force. The coupling of the pendulum to the works differs. One may even designate Calilei's escapement the more modern, since the pendulum at least swings free along part of its path. Considered from this point of view, it does not seem probable that Huygens had already been acquainted with Galileo's invention when he constructed his clock in 1657. It is probable that Huygens indeed knew the procedure for measuring short times with a pendulum driven by hand, a procedure already used by astronomers in Holland at that lime.

At almost the same time a further priority dispute was decided: Hooke contested Huygens' claim to the invention of the spring balance (Cf. Diehl-Patterson, 1952). However, this controversy shall not be presented here. It is probable that Huygens constructed a spring balance controlled by the same escapement mechanism as his pendulum clocks i.e., that of the spindle clocks, independently of Hooke; while Hooke may be credited with the construction of a spring balance with either anchor escapement or a prototype of the anchor escapement. Diehl-Patterson (1952, p. 279) closes her article, which also goes into the above-mentioned priority dispute, with the observation that, from 1657 on, the English were in contact with Huygens, and consequently an influence in both directions is very probable. According to Diehl-Patterson, at that time it was not a matter of developments being pushed forward by individual geniuses; rather, if anything, it was a chapter in the history of science in which collaboration of many men of learning prevailed (Cf. Diehl-Patterson, 1952, p. 316).

Huygens, moreover, occupied himself (as did Wren and Hooke) with a fully new form of clock: they anticipated a higher accuracy from a timing mechanism which revolved constantly than from one which oscillated back and forth.

Now theories of oscillation had already been developed in Oxford in 1655 and 1656. After Hooke, in his lecture 'The Inclining Pendulum' (Nov. 21, 1666) established geometrically a quantitative connection between the amplitude (i.e., the angle towards the vertical) and the 'urge' of the pendulum to return to the vertical (the *conatus* of returning to the center), the 'demonstrations of the circular pendulum' followed. Hooke knew that uniform circular motion can be composed from two superimposed sine waves and that consequently a circular pendulum oscillates harmonically. He constructed a clock with a constantly revolving timing device; it was presumably built in the year 1666.

A clock with conical pendulum more interesting because of its construction was "invented probably in 1659 or 1660 and built in 1667 and 1668" by Huygens. This conical pendulum clock of Huygens, which was indeed "extremely ingenious – yes, even could be called really beautiful", and which was also (Cf. Crommelin, 1950, p. 65) built and ran "fairly well" (*assez bien*, Huygens), could not, however, in principle function as a clock; something apparently unnoticed up to now!

$$T = 2\pi\sqrt{\frac{l \cdot \cos\alpha}{g}}$$

holds for the period of the conical pendulum (l = length, α = generating angle, g = acceleration of gravity). Huygens set himself the task of so constructing a clock that it automatically holds the period of oscillation constant. The conical pendulum meets this condition when $l \cdot \cos\alpha$ remains constant.

The following pendulum suspension meets the required condition: Two pendulum bobs are so fastened via thin cords to the tines *Z* of a swivel fork that, with an increase in the weights of *P* through centrifugal force, the cords slowly rise from the tines. The cogs *M* see to it that the pendulums do not fall behind the rotation of the fork.

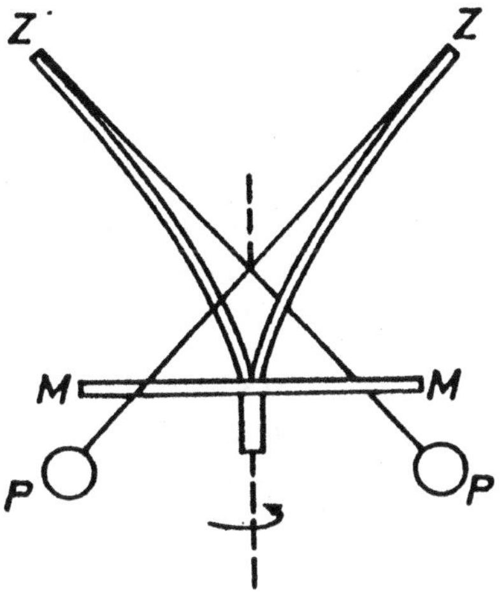

Fig. 25

Now if the tines are maintained in an evolute from a parabola, the pendulums, in their rotation and simultaneous rising, will describe a rotation paraboloid to which the cords are always perpendicular. Then

$$l \cdot \cos \alpha = p$$

holds[36], where p is the parameter of the parabola $y^2 = 2\,px$. Due to $T = 2\pi\sqrt{p/g}$, the period of oscillation T is independent of α. However, this only shows that a *free* conical pendulum has a frequency of its own independent of α.

In the ideal case this device cannot function as a clock when it is driven by a constant force (weight or spring), as the following dynamical reflection shows: supposing complete absence of friction, work is constantly supplied to the pendulum system. This work is absorbed and stored up as potential energy without change in the frequency of revolution only as long as the pendulums can ascend.

This process is limited by the finite length of the tines and cords. Thus, in the real case, the supplied work must be otherwise nullified, for example by the friction of the mechanism or by the resistence of the air on the pendulum bobs.

Thus it is not primarily the independent frequency of the conical pendulum which forces the escapement velocity on the device. Rather the driving force must be selected in such a way that it exactly balances the frictional forces when the pendulum revolves with its own frequency. In this case the pendulum can equalize the minimal variations in the ratio between the driving and frictional forces which disturb the equilibrium.

Since the conical pendulum is not joined with the movement by an escapement mechanism, but rather remains constantly coupled to it by cog wheels, the Huygens conical pendulum clock presents a mechanism whose synchronous movement rests on an equilibrium between drive and friction; because friction is so poorly controllable, this arrangement cannot prove a success.

Even if, with the achievements of Galilei, Huygens and Hooke, the story commences which was to give rise to a measuring technique of an accuracy to be found in no other area[37], the historical consideration of clock-building nevertheless breaks off here. If one disregards the progress which the technique of time-measurement has experienced in the 20th century – progress due on one hand to a highly developed technology and on the other to a board theoretical foundation – then the middle of the 17th century may count as that moment when the time concept of classical physics became prevalent. Thus a further look at subsequent clock-making to contribute to the understanding of the time conception of that day is no longer worthwhile.

At this point it is good to pause and to ask ourselves what presuppositions the builders of clocks may have tacitly made; to give an answer to this question here was precisley the reason for this short history of clock-making.

5. THE PRINCIPLES OF CLOCK CONSTRUCTION

The activity of constructing, building, and using clocks is not bound to [theoretical] propositions. What can be designated as the history of clock-making was, until Huygens, essentially a question of unreflective practice. Clock makers were no theorists and thus nothing has been written or preserved which gives information about what intentions were pursued in the clock-maker's art.

To gain such information there remains nothing other than to inspect the clocks themselves. Their construction principles allow us to draw conclusions after the fact about the time-conception of their inventors.

This way of seeking an insight into the tradition's extrascientific conception of time – above all, since it may be expected that such a conception will not differ essentially from today's, where surely all the old clocks could be utilized in still the same way today – harbors a special difficulty: the propositions which will be established about clocks contain terms of a modern technical language (explicitly introduced in this work) with all those distinctions whose availability for the old clock-makers is exactly the object of investigation. This dependence is to be taken into consideration in interpreting the principles of construction in terms of their presumable grounds or intentions.

In a distinction which cannot, however, be equally sharply observed through all the centuries of the period considered, we may distinguish two types of clocks. On the one hand, there are instruments which are supposed to produce as well as possible a uniform motion of constant velocity; on the other hand, there are those devices which repeat some event or other with as few differences as possible. The water clocks of Ktesibios or the conical pendulum clock of Huygens can be taken as examples of the first type; for the second type we may take the primitive scaleless water clock or the simple string pendulum which was used by astronomers to measure short times before the discovery of the pendulum clock.

The division into uniform and periodic clocks is less clearly applicable to the spindle and pendulum clock.

With the spindle clock the attention of the clock maker appears still to have been focussed primarily on the problem of so retarding the fall of a weight that it lasted sufficiently long and occurred with constant velocity. Thereby was achieved the constant turning of the hands. Whether or not the individual oscillations of the balance were now discernible was presumably of little interest. Rather the individual oscillations of the spindle clocks of that time were, as we can actually hear unaided, subject to considerable deviations (through inaccuracy of the crown-wheel). Only by averaging many oscillations can one speak of a constant period of oscillation.

However, with the pendulum clock, which depends on knowledge of the isochronism of pendulum oscillations, we can interpret the effort of the designer as an effort to realize increasingly better the indistinguishability of individual oscillations. The cycloidal pendulum of Huygens represents a special case of this endeavor which was supposed to make the isochronism directly independent of disturbing differences (strictly speaking certainly only of the amplitude)[38]. The mechanism which counts the oscillations did not offer constructively any further difficulties. Moreover when one now takes into account the fact that the pendulum clock of Galileo was not conceived for indicating the times of day, but above all was only to have been used as a *pulsilogium* or for astronomical observations – i.e., was to have enabled statements about the relationships of short durations –, then we may take it as even more probable that, from the invention of the pendulum on, the attention of the watchmakers was increasingly turned to achieving indistinguishable events. This path was trod with success until today. With modern quartz or cesium clocks this goal has been fulfilled to the most far-reaching extent. Modern spring dial clocks or possibly the indication of time of electronic clocks by means of counting tubes permits the uniformity of an actual constant motion to finally recede completely into the background.

In this connection it is astounding that no one has tried to give a linguistic formulation to such constructive goals. In no other area of practical life or science is the confidence of successfully working on an unreflective level so great as in time-measurement. In this fact we might see here a reason for the fact that into the 20th century chronometry was not developed into an exact theory, even though the example of geometry would have suggested this.

This unreflective confidence did not allow the problem of seeking theoretical foundations from the starting point of the practical effectiveness of time-measurement[39] to develop. Thus the foundation was not provided even after Einstein had destroyed the alleged unproblematic self-evidence of time-measurement.

If one – i.e., one who has become careful through the works of Einstein – restricts the problem of time-measurement to local time-measurement, then the careless conception of the uniformity of time appears to still possess full validity. Even at the level of higher abstraction, where talk of measurable time is already understood as invariant talk about artificially produced motions, the use of the words 'periodic' or 'uniform' on the basis of an exemplary or circular type of definition is still customary.

It seems plausible that the alleged lack of problems of time-measurement derives from the 'reliability' of astronomical events, particularly of the motion of the sun. Even this reliability, however, has never been earnestly questioned. (The knowledge that the earth rotation experiences deviations due to winds and tides, has in no way led to a search for an artificial time standard, the selection of which must be theoretically grounded. Clocks were built merely to replace the unpractical measurement of time based on astronomical time standards.) It may be left to other scientific disciplines to investigate to what extent mythic conceptions of the constellations, or perhaps physiological or other reasons, warrant the reference of our time sense and of our time-measurement to the motion of the sun. That it is just as obvious for the ancient Egyptian clock makers as it is for modern physicists is a triviality.

Now indeed there would be no objection if, as a consequence of this, physics were to remain based on our special situation, namely our living on earth. However, empirical physics has precisely elevated the disregard of this special situation to a program and this program forbids the use of standard events for time-measurement, which events are shown to be so only by way of example.

In pursuance of this program it can thus no longer be satisfactory to construct clocks which imitate astronomical events (i.e., to check clocks against sidereal days), whether this be in respect to their constant velocity or in respect to their indistinguishable repetition. Rather, exact empirical physics in its modern state requires that the terms of chronometry be fixed in a way going beyond the exemplary introduction given in Chapter III.

6. TIME THEORIES

In the preface to this chapter we said that we would select only such time theories which may be considered preliminary forms of a concept of time appearing again in the natural sciences. The selection of these theoretical accounts can be guided by criteria given in Chapter III, 1. Namely, since physics depends on determining ratios of durations, velocity ratios, forms of movement and the temporal order of events, it will suffice to look to discussions of the very question of the possibility of these propositions.

Further, to strive for new discoveries in historical studies or to search for still uninvestigated texts is less suited to the scope of this work than to undertake an interpretation of already known theories of time with the aid of a precise terminology. It is specifically the availability of specific distinctions and the knowledge of true propositions of time measurement which first permits a judgment of historical theories concerned with this subject.

There will be *one* theoretical account above all on which our attention will be focussed here. The Aristotelian theory of time, and this is precisely the thesis of the following discussion, has served as a foundation for many efforts to construct theories of time-measurement; however, in the further progress of history, it was increasingly garbled, until finally, even today, such a confusion with regard to time reigns that this historically first and conceptually distinguished theory is not understandable to many.

6.1. *Aristotle's Theory of Time.*

Aristotle's theory of time shows that physics' concept of time (disregarding the contributions of relativity theory) is already indebted to antiquity for its origin. Aristotle can rightly be taken as its creator. Other accounts neither fall under the stipulations given above, nor were they elaborated into theories (Cf. Wunderle, 1908). The definitions appearing in the *ὅροι* and attribtued to Plato: *χρόνος ἡλίου κίνησις μέτρον φορᾶς* (411, b, 3) remain outside of our consideration due to the doubtful authenticity of the *ὅροι* (Cf. Lesky, 1962, p. 556.)

Already the sufficiently known Aristotelian definition *τοῦτο γάρ ἐστιν ὁ χρόνος, ἀριθμὸς κινήσεως κατὰ τὸ πρότερον καὶ ὕστερον* (*Phys.* Δ 11, 219b, 1–2) indicates that Aristotle speaks about movements – and one might hope at least also about local movements of bodies –, and beyond that says that time shall be quantitatively (*ἀριθμός*) determinable.

There is such an abundance of works on the time theory of Aristotle that one might expect that we could rely on what has already been achieved. However one finds this expectation thoroughly disappointed.[40] These works[41] may be roughly divided into two groups, namely into 'ontologically' and methodologically oriented. The 'ontologically' inclined commentators expect from Aristotle primarily an answer to the question of the 'existence' of time. That he be interpreted in this direction is to be imputed to the Stagirite himself, since he asks, for example in *Phys.* Δ 2171, 30–35, whether or not time exists.[42] According to Wieland, the aporetic with regard to the existence of time is only preparatory. (Cf. Wieland, 1962, p. 323.) It may have its chief basis in the pronounced opposition of Aristotle to the Eleatic doctrine of time. The interpretations which place the accent above all on 'ontological questions' are far in the majority; this is because Greek and later, above all, scholastic commentaries on the Physics of Aristotle in turn gave rise to, commentators raised in the scholastic spirit. For obvious reasons they can hardly be of help to a reading which strives for methodological clarification of the Aristotelian text.

From the number of critical works – 'critical' simply understood as readable without recourse to esoteric linguistic customs – those of W. Wieland (Wieland, 1962) and G. Wunderle (Wunderle, 1908) stand out above all. In particular we may take Wieland's book as a well-founded presentation: thus our investigation can follow it to the greater extent. Of course some interpretation proposals at variance with Wieland will also prove to be suited to the text.

In his book Wieland comes to the conclusion that a concern with Aristotelian physics "in the light of the fundamental questions of modern sciences would be whorthwhile" (Wieland, 1962, p. 15) – a viewpoint which especially recommends Wieland's book for a systematic work.

In addition two theses explicitly taken by Wieland as a basis for his work speak to using this book; they are, for one, that Aristotelian physics is intelligible without recourse to metaphysics (Wieland, 1962, p. 15) and, for the other, that language is apostrophized as a central problem of philosophy in general and of Aristotelian interpretation in particular (Wieland, 1962, pp. 7–8), something which agrees with the tenor of our book – a proposal for standardizing speech about time.

Aristotle thoroughly deals with the way we gain knowledge about "that which exists by nature". About this let us say here only that for Aristotle the goal of research is a definitive representability of our knowledge in an axiomatically deductive form and "what interests him are the presuppositions which one must already have made if one would begin with a deduction of that sort". (Wieland, 1962, p. 53.)

This preliminary knowledge – "what is obvious and first known to us," – is always something poured together, (*συγκεχύμενον*), a complex knowledge. It is therefore valid to seek for the "*ἀρχαί*" and the "*αἰτίαι*", the principles and grounds. Here Aristotle is easily understood. "What is first known and obvious to us" is that (prescientific) knowledge which makes it possible for us to carry out life on a practical level (and it is, precisely in this respect, a presupposition for the possibility of philosophizing).

All the same an explanation of what Aristotle could have meant by "*ἀρχαί*" has its own difficulties. Arguments may be found equally for translating it as 'basic concepts' or as "basic propositions" in the sense of axioms. (Cf. Wieland, 1962, p. 52.) It does not contradict the text to think of '*ἀρχαί*' ('principles') as a temporary division of predicates introduced via examples into individual subject groups in order to arrive at first propositions on the basis of such a division. A (non-Aristotelian) example: predicates, which can be asserted or denied of events in ordinary language may be divided into local, temporal, causal, etc. To be able to standardize the use of these predicates, it helps to establish initially, (restricting ourselves to one group of predicates, – for example the temporal –), predicate rules for this group alone.

However we may interpret '*ἀρχαί*' in Aristotle (See here Section 14 of Wieland, 1962, '*Die Prinzipien als Topoi*'), we may say that it is a question of the beginning of human efforts to achieve reliable knowledge. In a phrasing open to more detailed suggestions of interpretation, let us therefore speak simply of 'beginnings'; in examining these 'beginnings' Aristotle is reflecting primarily on ordinary language. By means of a linguistic analysis in which he sees more than just a first step to a logical propaedeutic, he believes he can get a hold on that preliminary knowledge which one, *qua* speaking man, carries along *nolens volens* to every investigation (Wieland, 1962, p. 47).

Aristotle starts from the fact that there is change. Here change is a general concept for becoming and passing away, for qualitative and quantitative change as well as for local movements. The stipulation that the subject of physics is change of that which exists by nature already shows the central position of the concept of change in Aristotle. Since he discusses it under manifold aspects, it would be beneficial in endeavoring to understand his theory of time to remove some of these aspects from the spotlight. Thus, for example, according to traditional interpretation, a close relationship of the concept of motion to the modalities 'possible' and 'actual' comes to light in the discussion of the question of what motion is.

To be sure Wieland indicates that the frequently held view – movement is the 'transition' from the possible to the actual – depends on an inaccurate interpretation of the text; rather the "modal categories are applied to one another and in steps" (Wieland, 1962, p. 298, footnote 25). However, just how we are to think of the introduction of these modal categories in order for their iteration to make sense remains unexplained.

Indeed neither the question of what motion is nor the Aristotelian answer to it is a condition for understanding his theory of time; however, a clarified use of the word 'possible' must nevertheless be available when the concept of divisibility is introduced for the definition of the continuum.

Aristotelian modal logic offers, to be sure, as little help here as modern modal logic. 'Possible,' as a predicate of propositions, may be defined as 'not necessarily the negation of the proposition' and 'necessary' means as much as deducible from knowledge that one thinks has been given in propositional form. (See, for example, P. Lorenzen, 1955, p. 105f.)

Instead of using 'possible' as a predicate of propositions, it suffices, in the protophysical context, to introduce this word as a predicate for actions. An action is to be called 'possible for *x*' (*x* is a variable for men) if *x* knows an action schema.[43]

The proposition "An action *H* is possible for *x*" (where *x* is a specific man) is true when *x* actualizes the schema for *H* upon demand of a partner in dialogue.

Just as with the relation of the Aristotelian concept of movement to the modalities, so also what is treated of in *Phys.* A 7–9 as principles of the moved is likewise irrelevant for understanding the theory of time. There it is less change itself which is treated of than changeable objects and our knowledge about them.

The investigation begins where Aristotle speaks about movement (*Phys.* Γ, 1–3). That "local movement, change of place, already possesses some of the fundamental significance in the Aristotelian system which would later belong to it in physics" (Cf. Dijksterhuis, 1956) may be justified from the text.[44] We are therefore not doing violence to the text if, in the following, we always mean 'local movement of bodies' when we use 'movement'.

Movement can only be thought of as continuous (Wieland, 1963, p. 283) and "with the analysis of time, place and movement the structure of continuity is expressly presupposed". (Wieland, 1963, p. 279.) Therefore it is appropriate to first sketch Aristotle's theory of the continuum. (We should note that it would be incorrect to conclude from the lack of a formal presentation that the exactitude of this theory of the continuum is inferior to modern accounts.)

Aristotle quite consciously discussed the problem of the continuum within the scope of physics, because for him it was a question of structures of the experiencable (Cf. Wieland, 1963, p. 281); today, however, we are inclined to leave this question completely to the mathematicians. In spite of no end of parallels between the situation then and today – at that time two schools of opinion, represented by Zeno and Aristotle, stood opposed to each other, just as today the followers of Cantorian set theory oppose the followers of constructive mathematics – let the mathematically informed reader be warned not to tackle the Aristotelian theory of the continuum with modern distinctions in mind!

Aristotle gives two different definitions for "*συνεχές*": one time he defines it as "*τῶν τὰ ἔσχατα ἕν*" (whose outermost ends are one); another time as "*διαιρετόν εἰς αἰεὶ διαιρετά*" (divisible into an always further divisible).

The first definition appears together with the introduction of the terms '*ἁπτόμενον*' (touching) and '*ἐφεξῆς*' (contiguous), where the different possibilities of the position of extended magnitudes in respect to one another are discussed. These three terms are each two-placed, so that a two-place predicate must also be selected for the translation of '*συνεχές*'. Let us therefore suggest 'connected with' as a translation of the use of '*συνεχές*' as a two-place predicate, while use as a one-place predicate is translated as 'continuous'.[45]

Where continuity is spoken of, it is good Aristotelian usage to also always think of the object to which the predicate 'continuous' applies, since, according to Wieland, a '*Vergegenständlichung*' 'reification' (see Wieland,

1963, p. 230) of concepts such as place, time, movement, etc. is not appropriate to Aristotle. (By 'reification' is meant a grammatical change of propositions having predicates in adjectival form into materially equivalent propositions in which the adjective is replaced by a substantive – for example the transformation of 'A body is uniformly moved' to 'The movement of a body is uniform'. This substantivization of formerly adjectival predicates obscures their introduction and consequently offers an occasion for misunderstanding.)

Therefore 'continuity' is always 'the continuity of something continuous'. However, here we encounters the difficulty of having to state which object, to which the predicate 'continuous' is to apply, Aristotle intended.[46] In that Aristotle always thinks of methodically reconstructing scientific speech, the problem of whether it is length, duration or movement which is the primary continuous having the continuity of the others as a consequence now disappears. One can only introduce one word at a time, and nothing hinders us from choosing length as the subject on which the discussion of the problem of continuity will be based. This decision, independently of whether it is appropriate to Aristotle, has no direct effect on the interpretation of the definition of time.

Since a meaningful application of the predicate 'connected with' to two objects requires that boundaries (ἔσχατα) can be distinguished from non-boundaries in these objects – thus that they consist of distinguishable parts –, objects which have no parts can not be connected. This is an argument – corresponding proofs are given for 'touching' and 'contiguous' – against the Eleatic doctrine that a line can be thought of as a combination of points, where points are defined as objects having no parts. Just as a line is not constructible from points, so a movement is not constructible from thrusts (κινήματα), nor a duration from moments.

When Aristotle speaks of divisibility he always has a dividing authority in mind: man divides lengths, durations and motions.[47]

Precisely for this reason an understanding of 'divisible' is unproblematic. The possibility of continuous division is secured by knowledge of an action schema. To be sure the definition of 'divisible' or 'divisible for x' by 'knowledge of an action schema' can hardly be made out from the Aristotelian text. Rather it appears as though Aristotle would secure continuous divisibility in the object domain itself, as it were. In an example which recalls 'Achilles and the tortoise' he says that the quicker body divides time, the slower divides distance. (See *Phys. Z*, 233a; 7–8.) This aphoristic sentence stands for the following state of affairs: A body K_1, slower than a body K_2, travels a shorter way in a given time than K_2. The faster body K_2 travels this shorter way in a shorter time than K_1 requires for it.

In this shorter time the slower body K_1 travels an even shorter way than K_2 travels in the same time, etc.

To be sure, not only the fact that man is the dividing authority but also the fact that Aristotle mentions the procedure of carrying out divisions by setting marks (See *Phys. Z,* 263a; 23ff.), speaks against the conception of securing continuous divisibility in the object domain, as it were.

Confusion grows, when finally the three possible combinations for measurement are enumerated; namely measuring lengths and times, lengths and movements, and finally time and movements against each other (Cf. Wieland, 1963, pp. 212 and 283).

Indeed instead of regretting an "incomprehensibility of the essence of magnitude, time and motion" (Wieland, 1963, p. 291), an attempt to subordinate the intelligibleness of the Aristotelian assertions and to give a methodologically regimented interpretation that also fixes the word 'to measure' in particular recommends itself.

Agreeing with Wieland's opinion that the theory of the continuum precedes the definition of time, we will initially consider only the continuity of length.

A length '*μῆκος*', magnitude) is divisible into an always further divisible length, i.e., continuous, and its parts are connected. Man divides the length by setting marks on it. Ideation leads to a linguistic level on which it is possible to speak of the division of geometrical lines by boundaries and to require that the boundaries have no parts, i.e., are points.

Then, on the basis of such a geometrical continuum, talk of moving bodies and of dividing movements may be introduced, as was done in Chapter III, 2. Aristotle appears to have wished to present yet another possibility for dividing a movement: in that an observer utters '*νῦν*' twice, he has marked off an 'intermediate'.

Aristotle thereby extends distinctions which constitute the continuity of length to movement. The ('temporal') predicates known from ordinary language are made precise in their application by establishing rules strongly analogous to those for geometric terms. The boundaries, which are called '*τὰ νῦν*' (the 'moments'), and for which it is required that for themselves they have no parts, divide time. The parts are called durations.[48] Durations marked off by various moments are counted.[49] Thus time is the number (measure) of movement (divided into units of duration). Now Aristotle's definition indeed reads: time is the number of movement in respect to the earlier or later.

Whole generations of interpreters have tried their hand at the question of whether or not a circle lies in defining time by drawing upon the words 'earlier' and 'later'. As an obvious solution it has occasionally been recommended that '*ὕστερον*' and '*πρότερον*' be interpreted as referring to locations, something which indeed again raises the question of how a 'before' and 'after' are to be spatially differentiated other than by travelling a distance 'in time'.

Here one is faced with a pseudo-problem resulting from the unreflectiveness of ordinary language. As soon as the transformation of ordinary language into a restrained instrument suited to its purposes, such transformation resulting from a rational reconstruction, is comprehended as a meaningful possibility, this difficulty does not appear: 'earlier' and 'later' are introduced in examples (see p. 91).

No special feats of interpretation are needed to explain why Aristotle chose precisely these words. A duration is cut 'out of time' by two different boundaries. There is no word for moment in Greek and '*νῦν*' may not be used synonymously for moment (or point of time), since it only denotes the present moment, i.e., it stands in a definite relation to the actual speech situation (*Redesituation*). Logically viewed, '*νῦν*' is no predicate; rather it is a deictic expression, an 'indicator'.

Thus Aristotle chose the word pair 'earlier-later', which implies the divergence of boundaries determining a duration and which, in addition, may be understood without a theory of time.

Thus an interpretation of the Aristotelian definition of time has been presented which may be carried out non-circularly; nevertheless the interpretation requires the elucidation of two questions.

1. How is the term '*μετρεῖν*' (to measure) to be understood in Aristotle? (Besides the example already mentioned, that lengths and durations, lengths and movements and, finally, movements and durations are measured by each other, in the definition of time it occasionally says "time is the measure (*μέτρον*) of movement . . .". (For example in 220 b, 32, 221b, 22, among other places.)

2. Which conditions must a movement meet in order for it to measure time, since, according to Aristotle, movement can be faster or slower; time, however, does not (*Phys.* Δ, 218 b, 13–35)?

With the aid of definition of time, the use of 'to measure' may be conceived accordingly as the enumeration of units of duration; this it is always integers which appear as numbers of measurement. However, wishing to measure along a continuum with integers alone blessed Aristotle with critics, even among the Greek commentators, and among these Straton of Lampsakus in particular.[50]

Nevertheless several examples – say, the one cited above where the faster body divides time while the slower one divides distance; or the passage *Phys.* θ, 265 b, 8–11, where the movement of the heavens is designated as primary because it measures all others – allow the conjecture that a numerical determination of some magnitude or other is not always to be thought of with 'μετρεῖν'.

Here the modern use of 'to measure' impedes the approach to Aristotle. Before the reference to established units of measurement, common today, became prevalent, quantitative propositions of physics must have been written in the form of proportions.[51]

The representation of measurement-results by the equality of proportions, without using numbers [52] appears to come closest to the Aristotelian concept of measurement. 'To measure' in the area of dynamics means then marking off and comparing exact parts (lengths and durations) by drawing boundaries.

In any case, Aristotle proceeds in this manner when he speaks of the fact that lengths, movements and durations are measured by one another. To be sure, we have still not explained what the masurement of a movement is supposed to be. To think here of measuring the velocity of a movement is obvious; however this supposition is not supported by the text. Thus Wunderle suggests always interpreting a measurement of movement (or rest) as a measurement of duration;[53] this clearly leads to difficulties when movements and durations are to be measured by one another. This may be resolved if by 'duration' that duration of a process is meant which is compared with a movement, or more accurately a part of a distinguished movement.

Nothing can gloss over the fact that Aristotle did not express himself intelligbly in this respect. It appears as if no other possibility remains than to admit comparison of movements only with respect to distances and durations.

With the explanation of 'μετρεῖν' as 'to distinguish parts and to establish statements about the relations of durations and distances', an answer to the second question can be attempted – under what condition can a movement measure time?

The first part of the sentence 'change is always faster or slower, whereas time is not'[54] harbors no difficulty. In ordinary language the predicates 'faster' and 'slower' are used for observable changes. But what about the second part of the sentence?

Of two possible interpretations the version suggested again and again –

'time always passes uniformly fast' – is the incorrect one. Nowhere does Aristotle introduce the words 'fast' and 'slow' as predicates for time. In contrast to this, a reading which is thoroughly true to the text is the translation: 'Faster' and 'slower' can not be applied to 'time'. This sentence then has the rank of a predication rule for 'time', 'fast' and 'slow' and represents a requirement doing justice to ordinary usage. (Among others, passage 200 b, 1 supports this interpretation. There Aristotle expressly says, "One does not call time fast or slow, rather much, little, long or short". (*φανερὸν δε καὶ ὅτι ταχὺς μὲν καὶ βραδὺς οὐ λέγεται, πολὺς δὲ καὶ ὀλίγος καὶ μακρὸς καὶ βραχύς*.)

Whoever wishes to engage in talk of 'uniformly elapsing time' and does not thereby have in mind a subjective judgment of experiences, but rather intends scientific and thus situationally invariant statements, must either give a definition of time other than the Aristotelian one or at least establish rules as to how the predicates 'faster' and 'slower' are to be applied to this time.

Indeed, according to the Aristotelian text, not once does there emerge a sense of saying of time that it passes or moves itself. When we read in *Phys.* Δ 219 b, 23 that time follows (*ἀκολουθεῖ*) motion, this means nothing other than that duration increases with the distance travelled in a movement.[55]

According to Aristotle, in order for a movement to measure time it must run on a circular path. As the strongest argument for this he mentions the fact that rotary motion has neither beginning nor end, except when these are effected externally (*Phys. θ* 265 b, 14–16). The assertion that only a rotary motion can be uniform, while motion in a straight line is always non-uniform, is to be seen in the light of his theory of the natural places of the elements. Aristotle obviously is thinking of motions which occur in nature without the intervention of men. That an experimenter could artificially produce a standard movement is a thought which could only prevail with the overcoming of the Aristotelian natural philosophy by the founders of classical physics in the 17th century.

Thus it is not surprising when Aristotle marks the revolution of the heavens as the 'first' motion by which all others are measured. The methodological question, whether it must already be established what may be called a uniform rotary motion before the proposition that the revolution of the heavens is such a motion can be asserted did not arise for Aristotle because of 'metaphysical' reasons. However, it is exactly in metaphysical argumentation about the motion of the heavens that Aristotle reaches a point where his conception collides with the demand for methodological comprehensibility; thus our attempt to interpret Aristotle must break off here.

Besides uniformity, which nearly all commentators give at this point as the reason for setting apart the turning of the heavens – methodologically as uncritical as Aristotle himself – a cosmological basis may also be decisive. As the celestial globe is the "κοινός τόπος", the common place of all earthly things, so the revolution of the celestial globe is the "πρώτη φορά", the first motion, which produces the "χρόνος" as a time which encompasses all events within the celestial globe. Then, later, in the scholastic version, according to which 'time was' before the creation of the world, it was exactly this conception of Aristotle which was abandoned. In that, in particular, it was forgotten that talk about time was disguised talk about motion, time became independent. To be sure the motive for this change is not to be sought in physics, but rather in the intention to consistently interpret the theological account of creation.

Summed up in one sentence, the Aristotelian theory of time asserted: Time is measured against a circular movement by enumerating units of duration or by comparing (without the use of numbers) ratios of various parts of movements; this circular movement is, for 'metaphysical' reasons, uniform and eternal and is, analogously to geometric lengths, continuous.

One may agree with Wagner (Wagner, 1967) who says: "The consideration remains completely restricted to the needs of physics: time is discussed simply as the time of processes, as the measure of their duration".

We must reproach most works on the Aristotelian theory of time, with the exception of Wieland and Wunderle, with having overlooked precisely this fact.

If we compare the Aristotelian theory of time with the chronometry given in Chapter III, then the similarities of both accounts stand out. Disregarding the fact that the selection of the revolution of the heavens as a standard motion is not accounted for, rather at the most motivated in 'metaphysical' argumentation, the most one can reproach Aristotle for is that he has proceeded inconsequently in his terminology.

Here one must take into account that the individual books of the *Physics* were indeed not conceived as as connected works, but rather as a collection of individual lectures. Moreover in many passages, it is disputable, as the work of Torstrik makes clear, whether in the preserved texts one is dealing with the original or with later 'emendations'.

In summary, we may well assert that the theory of time of Aristotle is one of the great theoretical achievements of Greek antiquity. The conception of this program, which sets forth an important part of the presuppositions of an exact natural philosophy, is – in contrast to Aristotelian dynamics –

thoroughly suited, together with the theory of the continuum, to offer worthwhile help where it is a question of critically reflecting on the presuppositions of the natural sciences.

If, in later works, the Aristotelian theory of time occasionally gets a negative evaluation, this is due largely to the failure to recognize the fact that it is a piece of protophysics which is in question here. An interpreter will be encumbered if he approaches the Aristotelian text with his own detailed opinion of time, an opinion which not seldom includes all aspects of traditional speech about time.

Independent of such a judgment, the influence of the Aristotelian theory of time is again and again demonstrable in the history of the rise of the exact natural sciences.

If their remarks about time may be judged in one fell swoop, the Greek commentators on Aristotle's *Physics*' do not differ from him in the physical orientation of this theme. Simplicius as well as Straton of Lampsakus and Damaskios discussed preeminent questions of the continuum as well as questions of enumerating or measuring time. We must here forego searching for the first indications annoucing the gradual abandonment of this thematic orientation, an orientation which is still intelligible as a protophysical discussion.

Anyway, the turn – in modern terms – from discussing foundations of natural science to preferring other aspects, such as modes of time or the 'reality of time' in Augustine, took place. The two problems of time in the understanding of that period are briefly comprehended in the questions '*an sit*' and '*quid sit*'; later, in respect to deciding what time is, scholasticism subscribed largely to the Aristotelian definition. Nevertheless, with Augustine the question (always quite odd for methodological thinking) of time is moved into the foreground; it was treated there in connection with the modes of time. As a valuation of the role which Augustine's conception of time subsequently played, A. Maier writes: "The classic authority for this problem formulation for the Middle Ages was Augustine, with that famous chapter from the 11th book of the *Confessions*. The consideration, which above all became decisive for the following age is this: Time can be nothing real, since the parts out of which it consists are not real: The past is no more and the future is not yet and the present moment is only '*quia tendit non esse*'. And moreover, thus the scholastics completed this argument, the present moment is no actual, integral part of time, since time does not consist of *instantia*, just as in general a continuum does not consist of points. The scholastic thinkers, at least the vast majority of them, are fully aware of this".

To be sure, among the scholastics, motion (*motus*, in the most general sense of change) also had a primary status in comparison with the concept of time; this may be completely understood in the sense that talk of motions methodologically precedes talk of time. In general, however, the interest of scholars in this epoch was focussed on 'ontological' points of view. "The question which played a central role in the philosophy of the 13th and 14th centuries was formulated in the following: whether *tempus*, understood in the sense of the Aristotelian definition, had a reality outside the mind; whether it is *aliquid praeter animam*, whether an *esse in re extra* or only *in anima* alongs to it" (Maier, 1955, p. 43). It is surprising that this thematic orientation also appears to continue among the modern commentators on Augustine's theory of time, and with it the adherence to the discussion of the 'reality of time'. As a rule, then, the discussion of Augustine emphasizes the independence of this question in contrast to physics and its concept of time. Yet two facts are overlooked hereby: The discussion of the existence of time makes use of expressions such as 'divisible' or 'measurable' and 'part of time'. Even if Augustine makes only pure temporal propositions (Cf. p. 91) and no remarks on the measurability of time or clock-building by way of comparing movements, it nevertheless suffices for solving Augustine's paradox of time to elucidate the logical status of the words used by him. Moreover, we can find in Augustine a discussion – likewise restricted to pure temporal statements – of how time can be measured or, more exactly, which standard comes into question for time comparisons and for what reasons does it come into question. Thus, thematically, Augustine's theory of time also comes into consideration here; nevertheless we shall assert nothing about its historical efficacy in a development flowing into classical physics.

6.2. *Augustine's Theory of Time*

For Augustine, the problem of time reaches a peak in the XIth Book of the *Confessions* with the question: How can we measure time, since it certainly "is" not? The question of the existence of time is thus most closely connected with the question of its measurability.

Therefore, if the XIth book of the *Confessions* is interpreted here with a definite emphasis on scientific-theoretical interest, this is neither to suggest thereby that this book be read as a piece of theoretical philosophy or even as a methodological exposition, nor to disregard the fact the Augustine's discussion of time starts from an exegetical question about the account of creation (more precisely, *Genesis* 1:1.) and for Augustine is motivated accordingly. However, the treatment which Augustine's contribution to a philosophy of time has had appears to have been altogether one-sided[56] and overlooks – in any case in all those presentations cited in this book – the fact that the XIth book may be read and freely understood as an independent discussion. Admittedly, the emphasis on the scientific-theoretic aspect will leave some other viewpoints out of account which may be relevant for Augustine's exegetical problems. Above all, 'ontological' questions move into the background. Indeed, we may also state that the loss of the mentioned aspects affects primarily those aspects which first gain importance within the scope of far-reaching thematic preliminary decisions – preliminary decisions, to be sure, whose critical intelligibility has still to be shown.[57]

Our discussion begins after Augustine has designated time as created by God. Thus the state of the investigation reached at the beginning of the 14th chapter of the XIth book is significant here, because for Augustine time as oart of a nature created by God can thereby become the subject of a consideration within natural philosophy. It is this state of affairs which allows us to treat Augustine's discussion of time independently of theological or exegetical questions.[58]

Augustine begins, as though he were a clever student of Aristotle, with the things best known to him. (See Aristotle, *Physics* A 1, 184 a, 16.) We understand the word 'time' whenever we speak it or hear it from others.[59] The first approach to time, an approach theoretically not yet reflected upon, presents itself via the modes of time. The distinction of past, present and future is familiar to everyone. Everyone knows what it means for an event to be past or future. Such events do not take place now, are not present. Moreover – and this passage as well may be evaluated as going back to the preliminary understanding – time passes, i.e., what is present must later be past, or one would just have nothing to do with time.

Suspecting an incompatibility of this notion of time with it, Augustine first wonders at the fact, that we – and again one might add, in prescientific understanding – speak of longer and shorter time. (*Confessions*, XI, 14, 18 (p. 264) (In the following citations of the XIth book of the *Confessions* will only be in the form 14, 17 (p. 264.))) A past time span can only be long in the sense that it had a long endurance when it was present. On the other hand, every present time span, for example, the century in which we are living, may be subdivided; and only one part is present, the others being past or future (15, 19 (p. 265)). To this deliberation Augustine adds the conclusion that as long as the present has extension, it can be divided and can be classed among parts of the past or future; thus, strictly taken, it is not present.

However, as far as it has no extension – and Augustine finally restricts 'present' to this – it can also have no determinate extension, thus can not be measured or computed. Consequently the paradox yields itself to a pointed formulation: How can time be measured, since past, present and future 'are' not?

To resolve this paradox one must first hold to the fact that Augustine starts from a determination of the modes of time which determination is theoretically no longer problematic. In later passages he makes this distinction yet more precise by suggesting that one should speak of 'three times', namely the present of the past, the present of the present and the present of the future. Here again experienced time is obviously the starting point of the Augustinian discussion, since the past is present in momery, the present is present in direct perception, and the future in expectation (20, 26 (p. 269)).

Next it must be asked how 'to be' is to be understood in this context. To be sure, Augustine's talk of 'Being' in general shall not be investigated here (See here, for instance: Eckhard König, *Augustinus Philosophus*, Munich 1970, pp. 33–38); indeed, it will not be necessary at all to give a complete terminological definition of 'to be', where it appears as a full verb in the XIth book. Here it suffices fully, as will be shown, to keep a terminological restriction in mind. a restriction mentioned repeatedly by Augustine himself. (For example, 18, 23 (p. 268); 21, 23 (p. 270) and others.) The restriction is: What is not now, not present, 'is' not. That something is present thus represents a necessary condition for it 'to be' at all, whatever 'to be' may mean in this context. In modern logical terminology, the subsequent investigation, in reference to the use of 'being', follows the predicate rule: "What 'is', is 'now'", in symbols: $x \epsilon$ being $\Rightarrow x \epsilon$ now. Thereby "what is not 'now', 'is' not" is also logically valid, and this with the inclusion of corresponding synonyms such as 'present' for 'now' and 'existing' for 'being'. This terminological regulation certainly implies no synonymity of 'now' and 'being'.

Logically then, past and future have no 'being', since they were defined precisely as not being now. The argument for the fact that the present also has no 'being', can indeed not be similarly deduced from the terminological restriction of 'being' in such a simple manner. Indeed, Augustine makes an attempt to prove that the present 'is' not; but this attempt can be considered abortive and in this form plays no role in the further discussion. However let us mention it, if only for the sake of completeness: In order for the present to be time it must go over into the past, otherwise it would be eternity. But how can the present 'be', when "the reason why it *is*, is that it is not *to be*" (*causa, ut sit, illa est, quia non sit*)? (14, 19 (p. 264).) Thus we call the present 'time' only in view of the fact that it strives not to be. Augustine does not now add that therefore the present 'is' not. He also does not derive from this argument the conclusion that one cannot measure the present; rather he turns to a new reason for the fact that the present as well (besides, since they are not now, the non-existing past and future) can not be measured.

Just as the distinction of the modes of time in preliminary understanding appears to him to be given or at any rate at least to be theoretically unproblematic, so now Augustine proceeds, without further indications, from the fact that we can divide time. Here, then, he slips into his first serious error in assuming that by continuously dividing a present time (for example, one defined by the calendar) – the present century, the present year, the present month – he can finally come to an extensionless present; this is a mistake because the result of dividing an extended magnitude is always again extended.[60] The text shows unmistakably that Augustine violates this insight held to in the Aristotelian theory of the continuum. He argues further: were the present extended, then one could divide it and, except for one part, all parts would be past or future, thus precisely not present. Thus in a strict sense the present can only be like a point, which means nothing more than that it has no parts. However, if the present is not extended, then it can also have no determinable length. Thus, without yet speaking of the 'being' of the present, Augustine believes himself to have thereby given an argument for the fact that the present cannot be measured.

A look at the *Physics* of Aristotle shows that in this matter Augustine falls far short of Aristotle. The *νῦν*, the now, i.e., the present point in time, is according to Aristotle, no part of time; or in other words it does not measure time, since no duration can be composed out of indivisible points of time – no more than a line can be composed out of points.

On the other hand the remarks in Augustine on the use of the term 'to measure' are somewhat richer than in Aristotle. (p. 194 ff.) First there is the question of estimating durations; he mentions, for example, comparisons of spoken syllables of different length or comparisons of the circuit of the sun with the rotation of a potter's wheel. However, one might also think of measuring time with the help of sun and star dials as well as, certainly, with water clocks. A reference unmentioned by all commentators may be here that Augustine speaks about the division of the full day into 24 hours; thus he presupposes practically the measurability of time during the night as well.

Here measurement or estimation is an action which can only be performed on an available, present object. To be sure, events from memory can be estimated in their duration only so far as one brings them to mind; but the action takes place 'in the mind', and is thus not measurement in the narrower sense, i.e., it does not utilize measuring instruments or natural, observable processes. The same holds for the ability, which is closely connected with faculty of memory, to expect an event of a determinate duration. That the present cannot be measured with clocks has already been shown. Even after this sharpening of terminology the paradox remains: How can time be measured, since none of the 'three times' is measurable? To be sure the resolution of the paradox can not consist in an answer to the paradoxical question; rather it results from proving that the formulation of the question already contains a logical error.[61]

Whenever something is to be measured or estimated, i.e., wherever propositions about relations of magnitude are sought, everyone agrees that bounded magnitudes must either be imagined or indicated by establishing bounds. Even Augustine refers to the fact that one can only speak of the duration of an event if it is isolated or is conceived of as isolated. (24, 31 (p. 273.)) Now the incompatibility of such a measuring procedure with events on one hand and with the conceptual construction of the modes of time on the other derives solely from the fact that in this construction the present presumably appears as unextended. However, looked at closely. this construction already presupposes just that which is supposed to be proved impossible, namely a procedure involving isolated extended magnitudes.

In order to clearly point out the mistake in Augustine's argument, we must first keep in mind that 'extended' and 'divisible' are used synonymously, since, according to Augustine, every extended present event is also divisible; and whatever is divisible is, corresponding to the definition of antiquity, not a point, rather it is extended.

At the beginning of his discussion Augustine applies the predicate 'present' to extended events or durations (century, year, month, etc.). Without a doubt, his examples are all such that they presuppose the measurability of time in his sense, (i.e., Augustine makes no distinction between time-measurement and time-reckoning.) since the durations of which he speaks are nothing other than results of time-measurement. Thus we find a second error here; the measurability of time must be explicitly presupposed by Augustine for him to be able to argue for the proposition that time is not measurable.

As already mentioned, to assume an extensionless present as the result of repeated divisions is then an additional error. Nevertheless, if we ignore for the moment the fact that the division of an extended thing can only lead again to an extended thing, then, at the most, Augustine would have gained an argument for the proposition that the predicates 'present' and 'extended' or 'divisible' are contraries. If, in fact, Augustine wished to adhere to this strict use of 'present', then he ought not to have spoken of 'present' in connection with (certainly always extended) events. If time-measurement is the comparison of the duration of various events, then it can be concluded from this, that in the Augustinian terminology it is senseless to discuss the measurability of present duration, because 'present duration' is by definition a self-contradictory expression. However, that time cannot be measured does not follow from the fact that nothing present in the strict sense can be measured. The conclusion which Augustine should have drawn is: The question of the measurability of time has nothing to do with the modes of time in his sense.

Now nothing compels us to draw Augustine's incorrect conclusion along with him. On one hand, we can suggest that 'present' be used in the strict sense contrary to 'extended', 'divisible' and then also 'measurable'. Only in this respect is a point of time neither past nor future, rather present. On the other hand, we may still ask whether the continuous division of durations into increasingly smaller sections–each of which contains respectively the present points in time and each of which then, under precisely this condition, may still have 'now' predicated of it, allows the derivation of a proposition in regard to the measurability of time. Thus, if durations containing the present point of time are divided into shorter durations and if this division is again repeated on the part containing the present point of time, then one can speak of an arbitrarily short 'now', i.e., of a duration containing the present point of time. However, the result of such a series of divisions can never offer an argument for the proposition that time is not measurable. Producing the series of duration segments already makes use of the possibility of deciding that one duration is shorter than another. If here, too, comparing durations need not yet in principle be numerical, nevertheless estimating durations in regards to shorter and longer must already be possible.

Thus, in summary it can be said that Augustine makes three mistakes: (1) He overlooks the fact that the division of an extended magnitude only leads again to an extended magnitude. (2) After a correct argument for the proposition that the predicates 'present' and 'divisible' are contraries (if the present were divisible, then parts of the present would be past or future, thus not present), he violates this terminological rule by designating events as present. (3) In his argument for the proposition that time is not measurable, he presupposes precisely that durations can be divided and extended, i.e., that they can be measured.

At the same time it may have become plain that no 'ontological' properties of time of any sort are under discussion and that in no passage is there a concern with the question of whether time is 'actual' if the present is extensionless; rather it is clear that we must only establish the consistency of a terminology here or must indicate Augustine's violations of this consistency requirement. Thus Augustine's time paradox need concern philosophers no further. However this does not hold for the whole of the XIth book of the Confessions. Augustine appears to have sensed the precarious situation into which he landed with the question of the measurability of time based on terminological determinations of the modes of time. With this he takes a new tack, as it were, and discusses, without regard to the modality of time, how time can be measured. Up to now it appears to have been overlooked in the Augustinian literature that Augustine is earnestly coping here with a scientific-theoretical problem.[62]

He states anew what is of concern for him in his discussion: "My problem is to discover the fundamental nature of time and what power it has (*vim naturamque*). It is by time that we measure the movement of bodies. For example, we say that one movement takes twice the time that another takes." (23, 30 (p. 271)) His concern becomes even clearer in the following example. What is a day? Since the day is measured by the circuit of the sun, must one also then speak of a day if the sun completes its circuit in only one hour? His detailed question yet to be interpreted "whether a day is that movement itself [of the sun], the time needed for its completion, or a combination of both". (23, 30 (p. 271)) alludes to the search for a standard motion for time-measurement. Then unfortunately Augustine leaves the question – "what time is, for it is by time that we measure the course of the sun. If it travelled around the earth in a space of time equal to twelve hours, we should say that it had completed its course in half the usual time." (23, 30 (p. 271)) as unanswered as the attendant question – how the duration of movement and rest of arbitrary bodies can be measured, since movement itself is only possible in time. And since finally he does not succeed in finding "an accurate means of measuring time" (*certa mensura temporis*) (26, 33 (p. 274)) he satisfies himself with the pseudo-answer: "It seems to me, then, that time is merely an extension, though of what it is an extension I do not know. I begin to wonder whether it is an extension of the mind (*animi*) itself".

Thus, although Augustine does not end his discussion with an answer to the self-posed question, his argument shows – and in particular, the central section 23, 30–24, 31 – that he, also here going beyond Aristotle, addresses himself to decisive problems in the given context. First of all, the following passage presents some difficulties in interpretation here: ". . . Since, then, a day is completed by the movement of the sun through its total orbit, from the time when it rises in the west until it again reaches the east, my question is whether a day (1) (My parentheses) is that movement itself, (2) the time needed for the completion, or (3) a combination of both.

(E 1) ("E" for elucidation.) If a day were the movement of the sun through a whole circuit, there would still be a day even if the sun completed its course in a space of time as short as one hour.

(E 2) If, on the other hand, a day were the length of time which the sun actually takes to complete its circuit, it would not be a day if the period between one sunrise and the next were as short as one hour. In this case the sun would have to circle the earth twenty-four times to make one day.

(E 3) If, according to the third hypothesis, the movement of the sun and the time it takes to complete its circuit together constituted a day, there would be a day neither if the sun travelled through its complete orbit in the space of one hour, nor if it stood still while as much time passed as the sun regularly takes to circle the earth between one morning and the next." (23, 30 (pp. 271–272))

To begin with, the following interpretation offers itself: in (1) the movement of the sun is assumed as the solely available time standard, thus a *unit* of duration is defined by the course of the sun and this without supplementary possibilities of correction; while in (2) clocks independent of the sun are available, so that the day is defined by means of the *duration* of 24 hours, measured perhaps with a water clock. Yet this interpretation cannot be correct here. (1) and (2) are contradictory in this interpretation and thus allow no case (3) in which (1) *and* (2) are supposed to hold.

Even the most literal interpretation possible, i.e., that one may only speak of a day when the duration (measurable by clocks) of 24 hours has passed and when the sun has moved as well, – at least this is the reading permitted by explanation (E 3) – leads to a gross absurdity. That reference is made at all to the course of the sun is due the fact that in daily life the course of the sun determines the day. However, neither the accelerated revolution of the sun around the earth, accelerated in regard to another time standard, nor the standstill of the sun taken by Augustine from a Biblical example (*Joshua* 10:12 ff.) are to be met with in daily life.

However, since in (3) the sun-independent measurement of 24 hours obviously has priority in determining the day, one must here think not only of the case described, in which a circuit of the sun takes only one hour, but also one in which it takes much longer than 24 hours – only so long as the sun does not stand still. Thus many 'days' could lie between two sunrises; at the most the movement of the sun would have to assure that the passage of time is observable quasi-objectively. But a relation between the course of the sun from one morning to the next and the delimitation of a day is fully abandoned. However, since Augustine confirmed shortly beforehand that stars and 'heavenly lights' (sun and moon) exist to indicate times such as days and years, the interpretation that a day is the duration of 24 hours and that the sun must move as well in order for one to be able to speak of time is not suited to the text.

However, what does Augustine mean when he asks in (1) whether the day is the movement of the sun itself? What role does the question of by what means the day is determined play in the wider context, since Augustine seems to again abandon it so quickly? ("But for the present I shall not inquire what it is that we call a day. I shall confine myself to asking what time is, for it is by time that we measure the course of the sun.") The discussion of the relationship of the course of the sun and determination of the day is found in an argument against the opinion of a 'learned man', namely the opinion "that time is nothing but the movement of the sun and the moon and the stars".

And the counter-argument consists, as becomes clear from passages cited above, in the proposition that, for example, a day is a definite duration which maintains its sense independently of the course of the sun, more exactly even if the sun should suddenly no longer move uniformly – a day is determined by a non-astronomical time standard. Thus the passage under interpretation can be understood as an inquiry into the various possibilities of making celestial phenomena the time standard, where 'day' is an arbitrarily selected example and where another duration could be substituted for it. Augustine's inquiry may be consistently interpreted if one takes modern insight into the presuppositions of time-measurement as a basis. Whether Augustine anticipated this in detail may be left untouched.

Augustine appears to have seen that a standard for time-measurement is required in a two-fold respect: first a distinguished *form of motion*, such as that of the uniformity of the sun's course, is needed and then a *standard unit* as a definite division of the uniform movement must be defined. Here, as mentioned above, the sun and its motion stand for an arbitrary astronomical, a natural, time standard and 'day' for an arbitrary, astronomically defined, unit of measure.

Thus when in (1) it says that the day is the movement of the sun itself, this means that we elevate the movement of the sun to the standard for a form of motion, in other words declare it a clock. The reference, by way of example, to a natural standard implies that no standard other than a natural one is at our disposal, i.e., a time-measurement independent of the sun becomes impossible and (E 1) is fulfilled: no matter how the sun moves around the earth, one circuit counts as a fully determined duration, even if it lasts only one hour. To be sure whether the last restrictive condition can even be formulated methodologically at this point is still open. When we have no time standard other than the sun, it seems, above all, senseless to say "even if the circuit of the sun lasts only one hour".

Here Augustine's argument shows a double-level which, because of the lack of supplementary explication, gives rise to misunderstanding. Nevertheless, this failure to elucidate both 'levels' on which time is talked about, may be easily made good subsequently. In discussing methods – on one level – Augustine unfortunately makes use of means which, in the strict sense, only become available after completing the discussion of methods; namely he uses talk of determinate durations and time-measurement as it is customary in everyday life – and this is the second level. Thus when Augustine, in the discussion of methods, imagines the case where the circuit of the sun lasts only one hour, he does this with reference to everyday practice – for which a methodological inquiry was as irrevelant as it is today.

How little justified the reproach of circularity in 23, 30 is, may among other things be seen from the fact that Augustine has recourse to everyday customary talk of time only in explanations (E 1) to (E 3); however he uses it neither to formulate the methodological questions (1) to (3) nor to formulate any sort of presuppositions for a construction of a theory of measurable time. His problem in this passage does not consist in explaining how one may talk of time in general; rather it consists in explaining how from unreflected everyday practice one can come to a theoretically founded time-measurement which could be called scientific in the modern sense. After the movement of the sun is considered as a standard for a form of motion in case (1), (2) may be understood as asking whether the circuit of the sun is standard for a *unit of time*. Thus here it is tacitly presupposed that a form of motion has already been indicated, that in other words, it is already meaningful to speak of the uniform passage of time. One already has, so to speak, an uncalibrated clock, whose 'dial' bears no numbers or dividing lines. Suggestion (2) implies then nothing further than calibrating such a clock (one might think concretely of a water clock) by the course of the sun, i.e., undertaking a division of the dial according to the course of the sun. (Here one ought not let himself become confused by the fact that the clocks in use by us are already calibrated, in the sense that a full revolution of the hourhand represents one unit. Rather one ought to think of the problem of finding a linear scale – in principle infinitely long – as it is applied to the reading-off of a rising water level.) If then it should be somehow noticed, for example, by means of the calibrated clock itself, that the circuit of the sun suddenly lasts only just one hour, then it be asserted, in accordance with the establishment of the unit, that a day is only complete after 24 circuits – and this is what (E 2) implies.

If one reads (1) and (2) in the suggested way, then the third proposal also makes sense: if the course of the sun is standard for a movement form *and* at the same time for a unit, do we then regard the sun as a calibrated clock? Here Augustine advances a new explanation (E 3) which is not simply the conjunction of (E 1) and (E 2), though question (3) was merely the conjunction of (1) and (2). The interpretation of (E 3) is complicated by the fact that Augustine obviously reverses the order of explanation corresponding to the possibilities described in (1) and (2). The first part of (E 3) – one could not speak of a day if the circuit of the sun were to last only one hour – undoubtedly corresponds to suggestion (2). Then it is a question whether the second part of (E 3) – a day could also not pass, if the sun were to stand still, and let this be for a period of 24 hours – now belongs to supposition (1), as is expected. In fact this reading fits: If suggestion (1) consists in declaring the movement of the sun to be a standard for a movement form, then time can only be measured if the sun has moved at all and consequently can not be measured when the sun does not move.

The interpretation profered here, which also has the advantage over others that it consistently incorporates question (3) and explation (E 3) into the context, allows us to conclude thereby that it must have been more or less clear to Augustine that for time-measurement a standard is required respectively both for a form of motion and for a unit of measurement. He presents this insight with the help of the question whether the circuit of the sun is the first, the second, or both.

The adjacent passage shows that the suggested interpretation has not been gained by inadmissible restriction to a single passage. Rather Augustine can ask from a critical distance whether we ought to decide in favor of an astronomical standard, whether such a standard be a standard of the form of movement or of the unit; in any case, he decides against it. If he already had recourse to the possibility of time-measurement independent of the sun in explaining questions (1) to (3), so now he explicitly asks how one would have to measure the duration of a circuit of the sun to establish, for instance, that it had taken place in half the usual time. We may only conclude from this that he thought possible a method of measuring time which is to be preferred to the utilization of sun dials.

Here Augustine has the advantage over Aristotle that he was not prejudiced, as Aristotle was, by opinions about the divinity of the heavenly bodies and therefore did not ascribe uniform movement to them. Thus with a meaningful vocabulary, independent of the course of the sun, he can imagine states of affairs which consist in the standstill, acceleration or retardation of solar movement. In addition, Augustine recognized that such fictions in regard to the revolution of the heavens presuppose another time standard. He consequently also speaks of the movement of arbitrary bodies. (24, 31 ff. (p. 273).)

To be sure, Augustine does not here penetrate further to answer questions which were deferred, because he restricts himself completely to the durations of movements of bodies but does not go into the form of movement or velocity.

Nevertheless, within the scope of the discussion of what *motus localis* is, some suggestions occur which are again recognized in similar form in classical physics. The history of these beginnings shall be described indirectly.

6.3. *Transition to Classical Physics.*

In the dispute about the division of intensity, which follows on Oresme's theory of '*configurationes intensionum*' (about which we have yet to speak), the problem of using the intensio-remissio schema for local movement defined as '*motus ad ubi*' made its appearance. As regards the '*ubi*', the Aristotelian notion that '*locus*' was "the inner surface of the surrounding body, the ultimum contentis" (Cf. Maier, 1950) was adhered to. By it the '*locus*' of the '*mobile*' always moved along with the '*mobile*' so that it thus becomes a problem how the '*motus ad ubi*' could then be an '*aliter et aliter se habere*' in respect to the 'category of place'; in short, how could there be local movement when the place moved along with the moving body.

In contrast to the two proposals for solving this difficulty recognized at that time, the '*tertia opinio*' rejected by Buridan is precisely worthy of mention here; this is because this opinion speaks of a '*spatium separatum*' and '*infinitum ultra caelum*', which '*secundum modum penetrationis*' is the place of 'the world' (here to be understood as the place of 'all bodies'). This conception of space, which was largely rejected because the '*opinio multorum*' or the '*opinio apud vulgares*' of space as an empty receptacle is traceable to the imperceptibility of air, is hardly to be distinguished from the conception of space in Newton's *Principia*.

An analogous conception of time likewise stems from the 14th century. According to it "time is not – as Aristotle would have it – merely the *mensura motus* (and in particular the measure of the most distant celestial movement or the movement itself)" rather it is "a time existing from eternity and before creation of the world" (Maier, 1950, p. 23, footnote 20). Its advocate was Geraldus Odonis.

We shall not assert here that Odonis was the first link in a chain finally leading to the *Principia* of Newton; however, one may certainly view the scholastic contributions to a conception of time as a background against which the 'absolute time' of Newton, as usually interpreted today, stands out much less markedly than it does against, say, modern natural scientific doctrines.

That Newton was not the first to incorporate 'absolute time' into his natural philosophy is known. His teacher, Isane Barrow, speaks of uniformly passing time without reference to the experience of time in a lecture translated by Newton in 1669: "*seu currant res, seu stent, seu dormiamus, sive vigilemus, aequo tenore tempus labitur*". (Cited from Dugas 1954, p. 340.) Even earlier than in Barrow, the remark appears in P. Gassendi (in similar formulation) that time always passes equally quickly independently of whether now something occurs or not: "*tempus, seu sit aliquid quod ipso duret, seu non sit; et seu quiescat, seu moveatur; et seu moveatur ocyus, seu segnius; eodem tenore labitur*"; already in 1644 in his replies to Descartes he writes: "*sive res sint, sive non sint, sive moveantur, sive quiescant, eodem semper tenore fluit tempus*". (Cited from Rochot, 1956/57.) The similarity to Barrow's formulation is striking. Whether Gassendi influenced Barrow remains open; his influence on Newton may be taken as certain.

With his remark, Gassendi broke through the substance-accident schema of scholasticism and demanded for space and time a status of its own. To be sure, it is astounding in this connection that, in spite of this conception clearly reminiscent of 'absolute time' in Newton, a reference to time-measurement with clocks occurs ("imo et possent interim fluxu clepshydrae aut alterius machinae horariae distinquere"), something which obviously was thus not considered as contradicting the concept of 'absolute time'. With Newton also, the question of how to talk about time within the scope of his physics is not answered merely with the verbal definition of 'absolute time';

rather in the more detailed presentation we also find 'relative, apparent and ordinary time' (*relativum apparens et vulgare*). It pays therefore to ask once more whether the whole complex, involving the Newtonian definitions of 'absolute space' and 'absolute time' must be subjected to a new attempt at interpretation, since the reading of the scholium in question conveys the impression that injustice has befallen Newtonian physics (particularly in the contrast of classical and relativistic physics found in physics textbooks).

Indeed before we undertake a new interpretation attempt of the concepts of space and time in Newton – an attempt oriented less towards philological-historical principles than toward methodological principles – we will return one more time briefly to a scholastic account.

Oresme invented a graphic method of dividing '*intensiones*' over a '*latitudo*', which today is interpreted for the most part as a graph of one magnitude (distributed as ordinate) in its dependence on another (which is drawn as abscissa). Such an interpretation can evaluate Oresme's theory of the '*configuratio intensionum*' as one of the beginnings of classical physics (via D. Soto to Galilei). In contrast to this methodologically correct, but historically free interpretation, stands the statement of A. Maier which relegates Oresme's account to the region of "fantastic metaphysical speculation". The graphic method was indeed propagated solely through a treatise of the Oresme scholar and Augustinian hermit Jacobus de S. Martino or de Napoli, and in this way became known to Descartes and Galilei. (See Maier, 1950, p. 21, footnotes 17, 18.)

If one tries to summarize, in one fell swoop, the historical inheritance taken over by Newton before he composed his *Principia*, the result is a whole bundle of opinions in part incompatible with, in part independent of, each other. From the Aristotelian theory of time there remains only a paraphrased answer to the question of what 'ontological' status time has, an answer paraphrased according to the rules of scholastic thought. There was no longer talk of the movement of real bodies, or, to give a sensualist inflection to the Aristotelian account, of the observation of changes producing time. On the other hand – to be sure not theoretically, but certainly practically – the difficulty of Aristotelian physics was resolved, namely the problem of assigning numbers of measurement to durations from a continuous time. The practical successes of Huygens as well as Galilei's unconcerned talk of arbitrarily small parts of time accompanied the first steps in the infinitesimal consideration of continuous magnitudes in Newton's own differential calculus.

These two most important components – first the opinion of what time is, secondly time-measurement – are continued respectively in Newton, or, so it seems, in 'absolute, true and mathematical time' and in 'relative, apparent and ordinary time'. Now the Newtonian propositions concerning absolute space or absolute time are neither empirical nor can they be understood as hypotheses for explaining any phenomenon whatever. (The well-known objection, that Newton's profession to imagine no hypotheses is contradicted by the definitions of absolute space and time, misses the mark as it is, since he certainly conceived hypotheses to explain the forces of gravity working over a distance.) On the other hand, the theoretical level of Newton lets it appear advisable to look for a further role which those questionable definitions may have played in the methodological construction of his physics; and even if we concede that the text hardly offers a clue.

A sympathetic reading of the scholium to the definitions (even placing the scholium *before* the axioms supports this conception) can presume methodological principles behind these statements; these principles – loosely put – determine the relation of real objects to ideal obejcts. 'Absolute space' or 'absolute time' are thus nothing more than methodological means – in modern terminology – to secure the invariance of physical laws with respect to displacements in perceptible space or measurable time. The formulation of a principle, which – this may be recognized in the last draft of the scholium – would be a dynamical requirement indeed encounters manifold difficulties. At least one of these difficulties is unresolvable, namely formulating a principle which would give a definition of inertial movement without already having spoken of space, time and mass. However, even this free interpretation does not arrive at a decision as to whether Newton wished to distinguish one single system of reference or rather wished to distinguish all objects moved uniformly along inertial paths.

Let the attempt to understand Newton's concepts of time and space here be no more than a suggestion to ponder afresh this piece of Newtonian natural philosophy, since up to now interpretation attempts have been undertaken for the most part from some theoretical position or other which contain obstructive prejudgments.

As much criticism as Newton's theory of space and time has experienced even until today, we may nevertheless assert that as least his theory of time still survives in a modified form today. To be sure, no one wishes any longer to speak of a time which passes "uniformly without referencea to any external object whatever"; however, introducing standard movements by way of example for the purpose of measuring time, an introduction which offers no methodological justification for its choice (say, of inertial movement, light propagation, or periodicity of light) is no less 'metaphysical' than Newton's 'absolute time' in ordinary understanding.

NOTES

[1] Cf. D. Hilbert, 'Axiomatisches Denken', *Math. Annalen* 78, (1918), pp. 405–415. esp. 405. The separation of terminological problems from determining the validity of propositions disappears in Hilbert in that he locates concept formation there in an axiomatic theory where "the facts of a . . . field of knowledge" are ordered. Thereby he overlooks that such facts exist only in the form of propositions, which, for their part, cannot be formed without a terminology.

[2] The criticism of the synthetic *a priori* already expressed by empiricists of the 19th century, above all, against neo-Kantians, entered from the very beginning into the philosophy of the Vienna Circle on one hand via Ernst Mach and Moritz Schlick and on the other via Ludwig Wittgenstein and Bertrand Russell. Cf. Stegmüller, 1965, Chapter IX.

[3] This suspicion of metaphysics is most clearly expressed in Popper's criterion of demarcation. Although his falsificationism itself rests on methodological norms and is consequently not falsifiable, a lack of falsifiability already implies an unscientific nature. Thereby, for physical theories, falsificationism can be maintained thoroughly if Popper's solution of the basis problem is replaced by the proposal of explicitly normalizing the terms occurring in the basis sentences. (Popper, 1971. See on the basis problem in Popper also: Oetjens, 1975).

[4] Bunge employs the term 'protophysics' similarly to constructivistic protophysics: "The body of ideas shared by every set of physical theories but investigated by none in particular falls into two classes: the set of logical and mathematical theories constituting the formalism of physical theories, and a quaint collection of nonformal yet generic and important principles and theories, which may be christened protophysics." (*loc. cit.* p. 85). He sees, however, neither the normative nor the operational character that even a protophysics according to his own criteria must have if it is not to remain a merely formal theory. Obviously, German works on protophysics remained unknown to Bunge at least up to publication of his *Foundations of Physics*; to these works belong not only an essay by P. Lorenzen that had already appeared in 1961 and which sharpened the Dinglerian account – the term 'protophysics' was also already suggested there –; older works of H. Dingler also satisfy the criteria of Bunge's difinitions of protophysics; Moreover the word 'protophysics' is old; it already appears in J. Jungius (1587–1657) (as I was hinted to by G. Wolters) and is also presently employed in very different contexts with different meanings.

[5] The algorithm for determining the greatest common divisor of two numbers given in the 7th book, theorem 1, of Euclid's elements goes back to the older process of ἀνταναίρεσις or ἀνθυφαίρεσις (continued subtraction) according to which the length ratio of two (commensurable) segments results through alternate subtraction of the smaller (of the smaller remnant) from the greater (modern: in continued fraction representation). Thus the 'unit' thereby is ascertained *ad hoc*, i.e., for each respective length comparison as the greatest common measure.

[6] Hempel, 1952, p. 68. The discrepancy between the fact that coefficients of measure are always rational, but that physical theories, however, work with real numbers is only one of many reasons that have led to giving up the program of reducing the theoretical language of physics to observation language. In connection with the orientation of the theory-language account towards the model of formal-axiomatic theories, the fact that a methodologically correct solution of these problems can be sought other than by playing formal games on the metalinguistic level has receded into the background. The presently workable conception of the partial interpretability of theories is found in Toumela, 1973. Cf. also the detailed presentation which Stegmüller, 1973. Also there (in connection with Kuhn, Hanson, Feyerabend and Sneed), theories are spoken of only from a historical-descriptive view, without considering whether the problems solved by new accounts in the theory of science are not merely those of an older philosophy of science (Carnap, Popper) and still pass by physics as it is actually done.
[7] Expressly so, e.g., in Hilbert, 1918, or in Carnap, 1926.
[8] The technical questions on the selection of types of scales are found, e.g., in Suppes and Zinner, 1963, pp. 1–76 or Ellis, 1966, pp. 39–73.
[9] From Carnap's and Hempel's viewpoint it can here be replied that, according to the theory-language concept, interpretations should only be given for whole theories (here: geometry). It would not, however, be possible to establish a meaning for a single concept like 'length' in isolation. But if one considers, that as 'interpretations' of a formal geometry there come into question here propositions about artificially produced properties of measuring instruments, then additivity of segments does not appear as one of the last steps in defining 'length' and, at the same time, 'straight.' Rather it is the straight shape of segments whose lengths are to be compared, a property primarily enforced by the production of measuring instruments, which for its part logically implies the arithmetically analogous additivity of its parts.
[10] This example is explicitely discussed in Janich 1973.
[11] A slightly different solution to the problem can be found in E. Ströker. Still along the lines of analytic philosophy of science she assures that definitoric circles in science cannot be avoided in principle. But she adds that all measures to make our every-day-concepts more precise and unequivocal towards quantitative concepts can be justified only by prescientific experience (Ströker, 1973, 54). This however is no experience gained by the use of scientific instruments and therefore avoids logical or pragmatic circles.
[12] L. Schäfer, vicarious for many others, may be quoted: According to his diagnosis "P. Lorenzen and his pupils in Erlangen defend and develop Dinglers original intention, i.e., an operational variant of a transcendental 'last-foundation' (Letztbegründung)" which leads him to the classification of contructive philosophy of science as "conservative dogmatism of science" (konservativer Wissenschaftsdogmatismus) (Schäfer, 1974, p. 107). The Dinglerian fundamentalism however has always been explicitely refused by constructive philosophers. This fact has also been overlooked in the irrational criticism of protophysics by P. Feyerabend (Feyerabend 1978, p. 310).
[13] The 'three-disk' process offers an opportunity to produce flat surfaces on bodies without machine tools. For this three disks (roughly smoothed beforehand) are ground against each other in pairs until any two of the three disks fit on top of each other to the extent that they are slideable arbitrarily with respect to direction. This process, as a specimen of a pragmatically non-circular production of a geometric shape (the plane),

plays an important role in the constructive foundation of geometry. Analogously there are also given processes for the production of a right angle between planes. In contrast to the well-known process of constructing a right angle with compass and straight edge, the process employed in the constructive foundation of geometry makes no use of the congruence of line segments. Cf. Janich, 1976a.

[14] G. Böhme, in his review of the first German version of this book holds my criticism of Dingler to be unsound because "phase identity . . . (could) be ascertained without any recourse to periodicity". (Böhme, 1976, p. 103).

In fact, Dingler is concerned with 'phase identity', i.e., with the indistinguishability of individual phases of the repeated events. But one can speak exactly about indistinguishability only if the criteria are also stated, relative to which e.g., phases of events are indistinguishable. Böhme overlooks the fact that Dingler would at least have had to mention essential and unessential predicates in order to be able to require indistinguishability relative to the essential ones. Because, therefore, Dingler's definition of 'periodic' is incomplete, in his proposal he must proceed circulary to a realization process.

[15] 1968. Later relevant publications which already take into consideration the results of this book, but which do not essentially go beyond the *Logische Propädeutik* in the parts concerning protophysics are: P. Lorenzen, 1969, P. Lorenzen and O. Schwemmer, 1973. The latest development of Lorenzen's ideas about protophysics can be found in Lorenzen, 1984. (The reasons that the *Logische Propädeutik* and not, say, one of the important writings of Carnap, Hempel or another author is characterized here as 'the most compelling contribution' are developed in Chapter I, 2.)

[16] An impressive example for the dogmatism and irrationalism of an important kind of criticism can be seen in Kanitscheider 1971. He speaks about an experimental control of the propositions in protophysical geometry and chronometry – without any remark on the presuppositions of experiments without taking into account the normative character of protophysical sentences. Kanitscheider refers to Putnam 1962 although at that time no protophysical texts have been published, and recommends to study the principles of protophysics in Bunge, 1967, although the Bunge concept of protophysics and the constructivist approach have only the name protophysics in common. A strong affirmative attitude towards physics seems even to dispens with the effort to read the criticized texts and to understand the criticized positions.

[17] This objection is made by C. F. v. Weizsäcker who discusses the position of H. Dingler but seems to aim also at the to-day protophysics (v. Weizsäcker 1974). He says that his objection does not hinge on the empirical validity of relativity theory but only stresses the possibility of a supposition (Denkbarkeit einer Annahme) on the behavior of physical objects which should not be excluded by a certain methodology. A later passage however reveals itself the weakness of this argument: "A physical hypothesis is not impossible only by the reason that it seems phantastic relative to a methodology which is invented in order to avoid this hypothesis" ("Eine physikalische Hypothese ist nicht deshalb unmöglich, weil sie vom Standpunkt einer zum Zweck ihrer Vermeidung eingeführten Methodologie aus phantastisch erscheint", p. 63). v. Weizsäcker himself shows in that passage that he can hardly imagine of the "possibility of a supposition" (Denkbarkeit einer Annahme) different of a physical hypothesis. This hypothesis however needs terminology or at least some distinctions in advance. Besides it is historically

wrong to say that Dingler invented his methodology in order to avoid relativistic physical hypothesis.

[18] Only theorems (S 1) to (S 5) are relevant for the protophysical definition of uniform motion. The further parts of Section 3.1 have some bearing for a theory of time only with respect to the point that the theorems (S 6) to (S 8) are provable without time measurement.

[19] The following passage on the addition of velocities lead to a controversy between J. Pfarr and the author which is documented in R. Cohen and M. W. Wartofsky (eds.) 1984 and settled in Janich, 1984a.

[20] Thus the oldest discovered water clock bears the inscription (cited from Rend Taton (ed.), *Ancient and Medieval Science*, 1963): "Every figure is in its hour . . . to determine the hours of the night, when stars of the decan are invisible. Thus the correct hour of the sacrifice may always be observed". Nevertheless examples of other purposes for utilizing clocks are also known. Besides the use of water clocks in judicial practice, it is known from the Alexandrian age that the doctor Jerophilos used a pocket clepsydra for counting pulses.

[21] Cf. Mittelstrass, 1970. Mittelstrass traces the progress which Galileo's *scientia nova* represents over earlier projects back to the fact that Galilei united the traditions of the schools with a competence for physical questions (Paris and Oxford) with the 'tradition of the workshops'. In fact, the development of an exact empirical science is unthinkable without the combination of a protophysical theory with an experimental practice; this is valid even when the insight into this methodological distinction is not explicitly formulated.

[22] The datings of the technical literature vary quite widely. While R. J. Forbes and E. J. Dijksterhuis place the use of water clocks and the twelve-part division of the day or night at 2500 B.C., other authors date the division of the day into twelve hours first at 1400 B.C.. This discrepancy is presumably due to the fact that on one hand that Egyptian intermediate stage, in which the periods from dawn to dusk added up to ten equally long 'hours', is already cited as a twelve-part division of the day; while other authors speak of a division into twelve hours only then when the whole day, from sunrise to sunset, was divided into equally long segments. This is the division which also actually comes into use with measurement of time by clocks.

[23] J. Vercoutter remarks, quite correctly (in Taton, 1963, p. 41), that the Egyptian mathematics did not suffice for calculating the container shape that would have entailed a uniform sinking of the water level. Even the generating angle of the conical container, which in the clock found was given by an angle of inclination of the walls of 1:3 (according to Borchardt it must have amounted to 2:9, however, Cf. Lübke, A., 1958, p. 73), would have been arrived at empirically. In his remarks, Vercoutter appears to have overlooked the fact that, methodologically considered, the lack of mathematics would not prevent Egyptian clock makers from solving the problem. Namely, in order for the problem to be worked out mathematically at all, the dependence of the outflow velocity on the liquid's height must be known. Torricelli (1646) was the first one to establish the proposition that the outflow velocity is proportional to the square root of the height of the liquid above the drainage opening. However, Hero of Alexandria already knew that the outflow velocity is as great as if the liquid had gone the total height in free fall.

24 The inflow water clocks (the rising water level in a collecting vessel indicates the time) of Ktesibios were, according to Lübke (1958, p. 73), not the first of this form. In Egypt there seem to have already been inflow water clocks for which the problem of the scalar division of equal durations was solved.

25 In 1761 John Harrison, an English watchmaker, won the prize of £20,000 offered by the British government for the solution to the problem of determining longitude at sea within half a degree. His ship's clock lost only 15 seconds on a five-month voyage to Jamaica (see *Die Zeit*, from the series *'Life, Wunder der Wissenschaft'*, 1967, pp. 81–85).

26 *Walge=Walze* (cylinder). This clock consisted of a frame in which a cylinder hung on two cords wound around a shaft. As the cords unwound, the cylinder sank deeper and the shaft ends indicated the time in marks. In order that the cylinder would turn slowly, its interior was divided into individual chambers. Each of the dividing walls had a small hold through which a specific amount of water flowed respectively from one chamber into the next. (See Lübke, 1958, p. 78).

27 The collection of the Deutsches Museum in Munich contains many specimens of wooden wall-clocks and pocket watches.

28 Cf. Crombie, 1964, p. 186: "The early clocks were . . . mostly very large and the parts were made by a blacksmith. De Vick's clock was moved by a weight of 500 lbs. which fell 32 feet in 24 hours and had a striking weight of nearly three-quarters of a ton."

29 J. Drummond Robertson presents an exception, which assumes nevertheless a method of functioning in which the spoke whell swings back and forth. Indeed his explanation has two defects: he must assume that Villard incorrectly drew the direction in which the rope is wound on the shaft of the spoke wheel. Secondly, he must assume that the weight drawn smaller by Villard is the larger one. (Cf. Robertson, 1931.)

30 It is easily shown by Fig. 22 in the 3rd position that the validity of

$$r_W g > r_S G$$

does not change because of this slippage. Rather, since the friction of slippage is less than the static friction, the rotary momentum will be even smaller; thus $r_W g$ can effect a counterclockwise acceleration sufficient enough for spoke 1 to finally lose contact with the rope – a condition necessary for the oscillation capabilities of the system.

31 »Modus autem faciendi tale horologium talis esset, quod homo faceret unum circulum equalis ponderis ex omni parte secundum quod melius possibile esset. Postea quod appendatu (?) pondus plumbeum axi ipsius rote quod quidem pondus tali ter moveat rotam istam quod motus ille compleatur ab ortu solis usque ad occasum«, cited from Thorndike, 1914, 243.

32 The escapement mechanism of the spindle clock consists of a vertical crown wheel, in front of which stands a vertical shaft with two pallets set at approximately right angles. The shaft bears a cross bar (*foliot*) with two sliding weights for increasing the moment of inertia. The sliding of the weights served not only accurately to regulate the clock, but also to adjust it to the respective season, so far as the old twelve-part division of the solar day is taken as a basis.

Lübke gives an incorrect explanation of the functioning of the spindle clock: "The spindle . . . hung on a thread runner, . . . As the spindle rotated, the runner also twisted; since it thereby shortened itself, the spindle rose upwards releasing the balance until the next tooth of the balance wheel". It is not the raising and lowering of the spindle which releases the crown-wheel, but rather the spindle is accelerated in one direction until the pallet revolves completely out of the region of the teeth.

33 A detailed presentation of the events – as early as 1658, Huygens defended himself against the accusation that he had committed plagiarism, an accusation keenly supported by Hooke – is to be found in Robertson, 1931. Among new works on this theme let us mention in particular the article by Bedini, 1963.

34 »... io ho tal misurator del tempo, che se si fabricassero 4 o 6 di tali strumenti et si lasciassero scorrere troveremmo (in confermazione della lor giustezza) che i tempi da quelli misurati et monstrati, non solamente d'hora in hora, ma di giorno in giorno et di mese in mese non differirebbero tra di loro ne anco d'un minuto secondo d'hora, tanto uniformente caminano« (Ed. Naz. XVI, p. 467).

35 Prince Leopold de Medici, brother of the Grand Duke of Tuscany, Ferdinand II, founded the 'Academia del Cimento' together with Viviani, Torricelli, and others in 1657; he remained its president during the 10 years of its existence. In 1662 this circle, to which Galilei's estate was accessible, played an important role in the priority dispute. In a declaration from the 11th of August, Galilei's position as creator of the pendulum clock was stressed without, however, giving a description of the escapement. An enclosed drawing merely showed a closed case from which hung a pendulum. (On the role of this declaration see Robertson, 1931, p. 87ff.).

36 From the equation

$$y - y_0 = \frac{y_0}{p}(x - x_0)$$

(normals to the parabola $y^2 = 2px$ through the point P with the coordinates x_0, y_0) one obtains for $x(S) = p + x_0$, from which the projection QS of the normal piece PS to

$$QS = P$$

results.

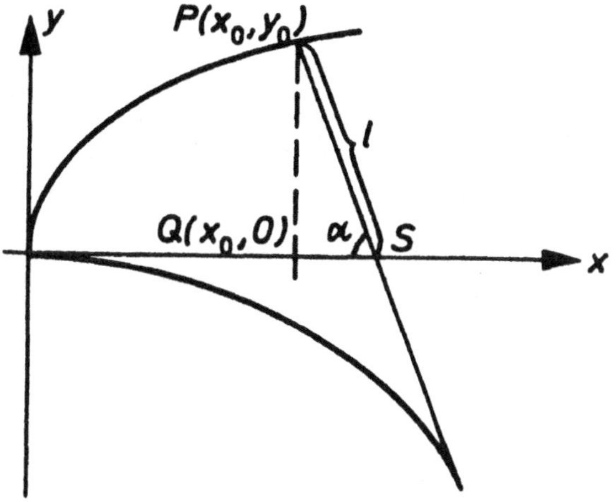

Fig. 26

[37] "Time can, in fact, be measured at least a thousand times more accurately than a length, a temperature or any other physical magnitude". (*Die Zeit*, p. 101.)

[38] Huygens judged his invention as the solution to the problem of constructing a correctly running clock: »nulla propter haec inaequalitas aut motus retardatio timenda erit, semperque aut recte tempus metictur aut omnino non metietur.« (Horologium oscillatorium. 1673, p. 6.)

This passage, moreover, allows one to recognize how unproblematic it was for Huygens to speak of the fact that "time is correctly measured" without also asking even once by what means 'true time' is to be distinguished.

[39] A compelling example of how clocks were constructed indifferently to theoretical justification is offered by Huygens. He says, in the context of a description of the cycloidal pendulum "... aut recte tempus metietur aut ominino non metietur ..." Without noticing the logical circle he establishes this assertion by means of the suggestion that no retardation (*retardatio*) or unevenness in running (*inaequalitas*) is to be feared with the cycloidal pendulum, while "a sure and equal measure of time ... does not inhere ... in nature". In the simple pendulum. "mensura enim temporis certa atque aequalis pendulo simplici natura non inerat", Huygens, 1673, I–II.

[40] So Torstrik, 1867, p. 480, reviews the achievement of the Greek commentators: The Greek exegetes also had no luck. They paraphrased individual expressions, changed the construction, and carried on a kaleidoscopic game with the concepts without having understood the matter. [Die griechischen Exegeten haben auch kein Glück gehabt. Sie umschreiben die einzelnen Ausdrücke, ändern die Construction und treiben mit den Begriffen ein kaleidoskopisches Spiel, ohne die Sache verstanden zu haben.]"

[41] P. F. Conen thinks the commentaries on Aristotle's theory of time should be divided thusly (Conen, 1965, p. 1.)

[42] Moreau, in particular, makes much of this. (Cf. Moreau, 1948, p. 59.)

[43] This terminological proposal, presented here independently of Aristotle, appears compatible with the Aristotelian concept of motion in its close connection to the concept-pair *δύναμις-ἐντελέχεια*. See C. F. v. Weiszäcker, 1967.

[44] According to Wunderle, *Phys*. VIII 7 in particular, shows the distinguished role of local movement. (See Wunderle, 1908, p. 47.)

Besides the examples which Aristotle uses in Z 1 and 2 and in other places, one may see plainly in 265b 17, as well, an explicit indication that local movement is to serve for the explanation of other changes and therefore ought to have methodological priority.

[45] Wieland, somewhat unfortunately, calls the first definition the "not yet proper" (p. 283), while the second is "final". He contrasts the "relational concept" of definition 1 with a "concept of quality" in definition 2 and distinguishes them grammatically according to number: in the plural one recognizes the relational, in the singular the concept of quality. However, to accord only a provisional or preparatory status to the use of *συνεχές* (connected with) as a two-place predicate and to then designate the one-place term 'continuous' as the goal of the whole consideration does not recommend itself; and, what is more, even then when (as Wieland thinks) the second definition takes into account "both aspects" (p. 285). A deviation in which the definition of 'continuous' would necessitate the already established use of 'connected', appears neither in Aristotle nor in Wieland. Rather using both terms together is in order.

[46] For Wunderle spatial extension is 'naturally' the primary. "A path is in itself a continuum, the movement as a consequence of the path it traverses, time finally through

movement". To support this assertion he refers to be sure, to the *Metaphysics* (*IV*. 13.1020 a 31) (see Wunderle, 1908, p. 54). Even though Wunderle leaves open what the phrase "a path is a continuum 'in itself' " is supposed to mean, one can nevertheless grasp its sense from his proposed sequence, if the continuity of the path is definitionally secured (switching over to geometry.)

Wieland, on the other hand, suggests that the continuity of motion, length and time can only be understood jointly. "Continuity is . . . a structure which still pervades those three forms of experience in curious ways. Magnitude as well as movement and time are continuous; however continuity is no 'property' which would belong to each of the three forms of experience while at the same time being independent of the other two forms of experience." (Wieland, 1963, p. 282.)

47 See Wieland, 1963, p. 301, where Wieland indicates that 'we divide' and 'we fix boundaries' are to be found frequently in the text.

When in citation above the talk is about a body dividing a distance or time, we can suppose that Aristotle is merely appealing to that procedure by means of which man divides.

48 Use of the duration here, is taken to be justified by Wunderle when he says: ". . . duration counts among the essential basic concepts of the Aristotelian theory of time and can in no way be eliminated from the same, however sparse the Stagirite's own indications way be". »... daß die Dauer zu den wesentlichen Grundbegriffen der aristotelischen Zeittheorie zählt und keineswegs aus derselben ausgeschaltet werden kann, so spärlich auch das Stagiriten eigene Andeutungen darüber scin mögen.« (Wunderle, 1908, p. 52.)

49 Wieland, in a thorough consideration (pp. 317–22), shows that with '*ἀριθμός*' one has to think of a number of units. Thus the measure numbers could only be natural numbers in Aristotle.

50 See here Wunderle 1908., p. 154. "Already in antiquity, Strato of Lampsakus, the student and disciple of Theophrastus, called attention to this. He emphatically opposed the use of the concept of number to define time. In totally rejecting the same he may indeed have gone a little too far, since number has, after all, proven itself to be significant in manifold ways for the concept of time; therein, however, he undoubtedly touched a sore spot of Aristotelian time theory in that, as Simplicius reports, he asserted as a chief reason for his opposition to the Aristotelian definition of time as the number of movement: *διότι ὁ μὲν ἀριθμὸς, διωρυσμένον ποσὸν, ἡ δὲ κίνησις καὶ ὁ χρόνος συνεχής, τὸ δὲ συνεχὲς οὐκ ἀριθμητὸν*.

Consequently [in Strato's own conception of time] sole dominion fell to the concept of measure and naturally from such a viewpoint the concept of *duration* also suggested itself more strongly than in the view of the Stagirite, whose remarks show a deficiency precisely in this respect.

51 Thus, to give an example, (anachronistic, to be sure) one would have written the equation defining force – $F = m\,a$, (m = mass, a = acceleration) – as $F_1/F_2 = m\,a_1/a_2$ before the introduction of units of measure.

52 Logically it is a description, something that becomes clear in the verbal formulation of the example in footnote 51:

Two forces working on the same body stand in the same proportion as those accelerations which they produce.

The representation of quantitative relations by means of proportions largely follows Euclid's theory of proportions; above all in the 14th century, as the '*calculationes*' came generally into fashion, a type of algebra for concepts taken to be quantitatively determinable was cultivated.

[53] Wunderle, 1908, p. 52: "This illustration further justifies assuming duration of both as the subject in all places where the talk is of a measurement of movement or of rest." [»Diese Erläuterung . . . rechtfertigt es ferner, an allen Stellen, wo von einer Messung der Bewegung oder Ruhe die Rede ist, als Gegenstand die Dauer der beiden anzunehmen.«]

[54] Phys. A, 218b, 13–15. *ἐτι δὲ μεταβολὴ μὲν ἐστιν θάττων καὶ βραδυτέρα, χρόνος δ'οὐκ ἐστιν.*

[55] According to the presentation by Dijksterhuis, even Damaskios appears to have understood the Stagirite thusly: ". . . if during a movement a sand-glass or water clock is running, then a number belongs to every phase of the movement, which number indicates the amount of water or sand which has run out since the beginning of the movement; this number is called time, . . ."

[56] Lechner merely comes to the following result which hardly goes beyond Augustine or elucidates him: "Thus the question of what time is leads to contradiction. But is not time thus nevertheless spoken of in that it manifests the contradication of its non-being Something (. . .) and gives as an answer to the question of its essence the statement that it eludes the question of its essence?" (1964, p. 133). Guitton restricts himself to a more psychological or historical-philosophical approach to Augustine's discussion of time, which discussion he painstakingly places into the context of the conceptions of time held by Augustine's predecessors. (Guitton, 1933).

Boros adds to Guitton's book, where it is a question "of discovering Augustine's coherent view behind the protean appearances of temporality." To be sure, he discusses, in a chapter about the 'metaphysics of temporality', an 'ontological' and a psychological question and examines questions of Augustine's conception of history; however, he leaves unmentioned Augustine's contribution to the question of the measurability of time (Boros, 1954).

Lampey attempts to interpret Augustine "phenomenologically-ontologically" (p. 22). He wishes "to understand . . . being by itself". In this interpretation time presents itself "to the watching subject as the contemplated object" (p. 25). In proceeding as an interpreter from such a viewpoint, it certainly promises to be of advantage that this position still partly stands in the Augustinian tradition. However, Lampey also never goes beyond the scholastic tradition and cannot make Augustine intelligible to that reader who first asks for reasons why he should think himself into this tradition (Lampey, 1960).

Eibl appears to be the most unbiased in approaching the Augustinian text in his chapter 'Das Zeitproblem' (The Problem of Time). However he fully misses Augustine's intention when he contrasts an actual, extended present with one given "only in abstraction" which is punctiform as well as enjoying equal rights with the actual one (Eibl, 1923, p. 337).

[57] Let v. Del Negro, 1966, pp. 301–312, be mentioned as examplary of such theoretical prejudgments in favor of 'ontological' formulations and the difficulties resulting from them. This "attempt to again rehabilitate the reality of the immanent-ontological which appears undermined in a dangerous manner by the time paradox" overlooks the fact that a punctiform, i.e., indivisible Now is the result of a conceptual construction and that consequently only logical or terminological demands can be placed on such a

construction. To formulate questions of the 'reality' of the present or of time already falls short in the given context; because only he can speak of an indivisible present who already knows in what manner time is divided and thus bears in mind that he has thereby theoretically acceded to human actions or instructions for action.

In one of the newest works (Schmitt, 1967) – a work according to the introduction "based on a philosophical rather than a theological approach" – the theoretical prejudgment, is itself not discussed, ("However, What-is (*Was-sein*), essence, only exists for that to which Being belongs" p. 42) works itself out in the statement that "the inquiry into the essence of time . . . (must) be taken back to inquiry into the being of time." Thus the answer to the question "Does being, however, belong to time?" (p. 42) remains unsatisfactory; for which reason a further thematic limitation is supposed to offer elucidation: ". . . The same must be said of time. Its investigation as well is not concerned latterly with time as time, rather represents . . . a step along Augustine's path of knowledge to God". p. 65.

58 See here U. Duchrow, who, in reference to the intellectual-historical background of the Confessions, opposes the one-sided emphasis on the psychological aspect of Augustine's discussion of time; he also refers to the fact "that the constellations are demythologized in the Hebraic-Christian tradition and have nothing to do with angels and not a thing to do with God's essence." p. 283 (Duchrow, 1966, pp. 267–288).

59 See *Confessions, XI*, 14, 17 (p. 264 d) [For Augustine, see the Penguin Books edition of the *Confessions*, translated by R. S. Pine-Coffin. The pages numbers in parentheses refer to this edition. -Tr.]

60 The violation of the Aristotelian insight – that it is man who divides time and that consequently, due to the finitude of all actions, a point of time can never be obtained by division – is also repeated in newer works; thus, for example, in the dissertation of Gutwengler. In this work which contrasts a 'formal conception of time' (questions of time measurement) with a 'material structure of time' ("accessibility of time in consciousness") and which asserts "Time suffers damage by means of this view" (i.e., by means of the "spatialization of time, particularly as classical physics represents it"), it is stated: "It is still always possible to subdivide a period of time into smaller parts until a point is left as the limit of divisibility". (Gutwengler 1955, p. 42).

61 R. Suter also takes this path. He remarks correctly that Augustine's difficulties stem not lastly from his own statement of his concept of time ("Augustine's picture of time is largely responsible for his difficulties", p. 387). To be sure, Suter overdoes the analytic method when he reproaches Augustine for simply overlooking the fact that words such as 'present', 'enduring' and 'current' may be used in a broader and narrower sense, namely as predicates for longer and shorter durations. In this passage Augustine is evidently only concerned with proving that – in modern terms – 'present' and 'divisible' are incompatible in a terminologically strict sense. In no case does Augustine's insight lead, as Suter asserts, to a distorted use of the word 'present'; where this word is permitted by an exact terminological definition for the purpose of scientific speech, and can also be further applied meaningfully to durations in ordinary language. Suter's analytic interpretation suffers precisely in failing to make this distinction, namely the distinction between non-standardized ordinary language and terminologically explicitly standardized scientific language. (Suter, p. 387–394.)

68 Of all the works cited in this account, that of J. M. Quinn is the only one which comments on the important passage 23.30. Nevertheless this commentator misses his

target in two ways: First he overlooks the fact that, in the passage concerned, Augustine does not discuss the question 'What is time?; rather he is explicitly seeking an answer to the question 'How do we measure time?. Thus in no way does Augustine speak of the circuit of the sun here (23.30) in order to show that the movement of the sun is also not time (Quinn's interpretation – a completely superfluous intention at this place, since Augustine has just shown that time is not the movement of the heavenly bodies, expressly mentioning among these the sun (23.29). Secondly, in no part of the text is a *reductio* such as that constructed by Quinn (one hour is equally 24 hours) to be discovered. Quinn's interpretation proves to be entirely unsuitable, since the conjunction (*utrumque*) of both possibilities mentioned by Augustine – a day is the movement of the sun and a day is the duration of a circuit of the sun – is not possible in it. Then, somewhat perplexed, Quinn also remarks on this third proposal formed by the conjunction of the first two possibilities: On the third hypothesis, a day ceases to exist altogether at certain junctures. It is not very illuminating to suggest as a solution just where Augustine is concerned precisely with the question 'How do we measure time?' that: "The truth of the matter is the other way around. It is not – solar motion that measures time; rather it is time that measures solar motion" (p. 85ff). (Quinn, 1969, pp. 75–127).

REFERENCES

Andresen, C., Erbse, H., Gigon, O., Schefold, K., Stroheksr, F. K., Zinn, E. (eds.): *Lexikon der alten Welt*, Zürich-Stuttgart, 1965.

Angelelli, I. (ed.): *G. Frege, Kleine Schriften*, Darmstadt, 1967.

Ascheberg, R., Herrgott, G., Krausser, P.: 'Zur konstruktivistischen Protophysik der Zeit', in: *Zeitschrift für allgemeine Wissenschaftstheorie IX*, 1978, 112–133.

Bassermann-Jordan, E. v.: *Uhren*, Braunschweig, 1961.

Bedini, S. A.: »*Galileo Galilei and Time Measurement*«, in: *Physis* V, Z, 1963.

Böhme, G. (ed.): *Protophysik, Reihe Theorie-Diskussion*, Frankfurt, 1976.

Boros, S. J., L.: *Das Problem der Zeitlichkeit bei Augustinus*, Diss. München, 1954.

Bridgman, P. W.: 1952. *The Nature of Some of Our Physical Concepts*, New York.

Büchel, W.: 'Zur »Protophysik« von Raum und Zeit', *Philosophia Naturalis* 12, 1970, 261–281.

Bunge, M.: *Foundations of Physics*, Berlin/Heidelberg/New York, 1967.

Carnap, R.: *Physikalische Begriffsbildung*, Karlsruhe, 1926.

Carnap, R: 'Über Protokollsätze', in: *Erkenntnis* 3, 1932.

Carnap, R.: *Philosophical Foundations of Physics*, New York, 1966.

Cohen, R. S. and Wartofsky, M. W., *Physical Sciences and History of Physics* D. Reidel, Dordrecht Boston Lancaster, 1984.

Conen, P. F.: 'Die Zeittheorie des Aristoteles', *Zetemata* 35, München, 1964.

Crombie, A. C., *Augustine to Galileo. The History of Science A. D. 400–1650, I–II*, London, 1952.

Crommelin, C. A.: 'Die Uhren von Christian Huygens', in: *Endeavour* IX, 34, 1950.

Diederich, W.: *Konventionalität in der Physik*, Berlin, 1974.

Diehl-Patterson, L.: 'Pendulum of Wren and Hooke', in: *Osiris* (1952), Bibliotheca Harvardiana 185.

Dijksterhuis, E. J.: Die Mechanisierung des Weltbildes, Berlin Göttingen Heidelberg 1956.

Dingler, H.: *Grundlagen der Physik*, Berlin/Leipzig, 1919, 1923.

Dingler, H.: *Physik und Hypothese*, Berlin/Leipzig, 1921.

Dingler, H.: *Die Methode der Physik*, München, 1938.

Dingler, H.: *Die Ergreifung des Wirklichen*, München, 1952.

Duchrow, U.: 'Der sogenannte psychologische Zeitbegriff Augustinus im Verhältnis zur physikalischen und geschichtlichen Zeit', *Zeitschrift für Theologie und Kirche* 63, 1966.

Dugas, R.: *La Mécanique au XII Siècle*, Neuchâtel, 1954.

Eibl, H.: *Augustin und die Patristik, Reihe* »Geschichte der Philosophie in Einzeldarstellungen«, München, 1923.

Ellis, B.: *Basic Concepts of Measurement*, Cambridge, 1966.

Essler, W. K.: *Wissenschaftstheorie I–III*, Freiburg/München 1970–1973.

Fertig, H.: *Modelltheorie der Messung*, Berlin, 1977.

Fertig H.: 'Wie objektiv ist die Physik? – Zur Diskussion um die Protophysik', in: *Philosophia Naturalis* 17, 1978, 31–55.

Feyerabend, P. K.: 'Die Wissenschaftstheorie–eine bisher unerforschte Form des Irrsinns?' in: P. K. Feyerabend *Der wissenschaftstheoretische Realismus und die Autorität der Wissenschaften*, Braunschweig, 1978.

Frege, G.: 'Über das Trägheitsgesetz', in: *Zeitschrift für Philosophie und philosophische Kritik* 98, 1891, 145–161.

Gabriel, G.: *Definitionen und Interessen*, Stuttgart-Bad Cannstatt, 1972.

Gethmann, C. F.: *Protologik. Untersuchungen zur formalen Pragmatik von Begründungsdiskursen*. Frankfurt, 1979.

Grieder, A.: 'Protophysik der Zeit und Relativitätstheorie', *Dialectica* 30, 1976, 145–159.

Guitton,. J.: *Le Temps et l'Eternite chez Plotin et Saint Augustin*, Paris 1933.

Gutwengler, E.: *Der Zeitbegriff bei Augustinus*, Diss. Wien 1955.

Heath, I., (ed.): *Euclids Elements I*, New York, 1956.

Helmholtz, H. v.: 'Zählen und Messen, erkenntnistheoretisch betrachtet', in: *Philosophische Aufsätze, E. Zeller zu seinem fünfzigjährigen Doktorjubiläum gewidmet*, Leipzig, 1887, 17–52.

Hempel, C. G.: *Fundamentals of Concept Formation in Empirical Science* (International Encyclopedia of Unified Science II, 7), Chicago, 1952.

Hempel, C. G.: *Grundzüge der Begriffsbildung in der empirischen Wissenschaft*, Düsseldorf, 1974.

Hilbert, D.: *Grundlagen der Geometrie*, Stuttgart, 1968.

Hilbert, D.: 'Axiomatisches Denken', in: *Mathemat. Annalen* 78, (1918), 405–415.

Hörz, P. and Wollgart, S. (eds.): *H. v. Helmholtz, Philosophische Vorträge und Aufsätze*, Berlin, 1971.

Hucklenbroich, P.: *Theorie des Erkenntnisfortschritts. Zum Verhältnis von Erfahrung und Methoden in den Naturwissenschaften*, Meisenheim, 1978.

Huygens, Ch.: *Horologium osillatorium*, Paris, 1673.

Inhetveen, R.: *Konstruktive Geometrie. Eine formentheoretische Begründung der euklidischen Geometrie*, Mannheim/Wien/Zürich, 1983.

Janich, P.: 'Eindeutigkeit, Konsistenz und methodische Ordnung: normative versus deskriptive Wissenschaftstheorie zur Physik', in F. Kambartel and J. Mittelstraß, (eds.), *Zum normativen Fundament der Wissenschaft*, Frankfurt, 1973.

Janich, P., Kambartel, F., and Mittelstraß, J.: *Wissenschaftstheorie als Wissenschaftskritik*, Frankfurt, 1974.

Janich, P.: 'Trägheitsgesetz und Inertialsystem. Zur Kritik G. Freges an der Definition L. Langes', in: Ch. Thiel, (ed.), *Frege und die moderne Grundlagenforschung*, Meisenheim, 1975.

Janich, P.: 'Zur Protophysik des Raumes', in: G. Böhme (ed.), *Protophysik*, Frankfurt, 1976a, pp. 83–130.

Janich, P.: 'Zur Kritik an der Protophysik', in: G. Böhme (ed.), *Protophysik*, Frankfurt 1976b, pp. 300–350.

Janich, P.: 'Das Maß der Masse', in: K. Lorenz (ed.) *Konstruktionen versus Positionen*, Berlin, New York, 1979, pp. 340–350.

Janich, P.: 'Ist Masse ein »theoretischer Begriff«?' in: *Zeitschrift für allgemeine Wissenschaftstheorie* VIII, (1977a), 302–314.

Janich, P.: 'Die Sprache der Physik und die Wirklichkeit der Naturwissenschaften', in: *Dialectica* **31**, (1977b) 301–312.

Janich, P.: 'Die Protophysik der Zeit und das Relativitätsprinzip. Erwiderung auf J. Pfarr', in: *Zeitschrift für allgemeine Wissenschaftstheorie* **IX/2**, 1978, 343–347.

Janich, P.: 'Physics – natural science or technology?' in: Krohn W., Layton E. T., and Weingart P. (eds.), *The Dynamics of Science and Technology*, Dordrecht, 1978, pp. 3–27.

Janich, P.: (ed.), *Wissenschaftstheorie und Wissenschaftsforschung*, München, 1981a.

Janich, P.: 'Natur und Handlung. Über die methodischen Grundlagen naturwissenschaftlicher Erfahrung', in: O. Schwemmer (ed.), *Vernunft, Handlung und Erfahrung*, München, 1981b, pp. 69–84.

Janich, P.: 'Newton ab omni naevo vindicatus', in: D. Mayr and G. Süßmann (eds.), *Space, Time and Mechanics*, D. Reidel, Dordrecht, 1983, pp. 225–240.

Janich, P.: 'Commentary on "Protophysics of Time and the Principle of Relativity" ', in: R. S. Cohen and M. W. Wartofsky (eds.), *Physical Sciences and History of Physics* (Boston Studies in the Philosophy of Science 82), Dordrecht/Boston/Lancaster, 1984a, pp. 191–198.

Janich, P.: 'Irrtum oder Methode? Zu den handlungstheoretischen Grundlagen, der Protophysik in Auseinandersetzung mit der Kritik A. Kamlahs', in: *Zeitschrift für allgemeine Wissenschaftstheorie* **XV**, 1984b, 122–141.

Janich, P.: (ed.), *Methodische Philosophie. Beiträge zum Begründungsproblem der exakten Wissenschaften in Auseinandersetzung mit Hugo Dingler*, Mannheim/Wien/Zürich, 1984c.

Janich, P.: 'Hugo Dingler, Die Protophysik und die spezielle Relativitätstheorie', in: P. Janich, 1984d, pp. 113–127.

Kambartel, F.: 'Frege und die axiomatische Methode. Zur Kritik mathematikhistorischer Legitimationsversuche der formalistischen Ideologie', in: Ch. Thiel (ed.), *Frege und die moderne Grundlagenforschung*, Meisenheim, 1975, pp. 77–89.

Kambartel, F.: 'Überlegungen zum pragmatischen und zum argumentativen Fundament der Logik', in: K. Lorenz (ed.), *Konstruktionen versus Positionen*, Berlin, New York, 1979, pp. 216–228.

Kamlah, A.: 'Zwei Interpretationen der geometrischen Homogenitätsprinzipien in der Protophysik', in: G. Böhme (ed.), *Protophysik*, Frankfurt, 1975, pp. 169–218.

Kamlah, A.: 'Zur Diskussion um die Protophysik', in: K. Lorenz (ed.), *Konstruktionen versus Positionen*, Berlin/New York, 1979, pp. 311–339.

Kamlah, A.: 'Methode oder Dogma? Eine Auseinandersetzung mit der zweiten Auflage von P. Janichs Protophysik der Zeit', in: *Zeitschrift für allgemeine Wissenschaftstheorie* **XII/1**, 1981, 138–162.

Kamlah W., and Lorenzen P.: *Logische Propädeutik*, 1967, 1973.

Kanitscheider, B.: *Geometrie und Wirklichkeit*, Berlin, 1971.

Kant, I.: *Kritik der reinen Vernunft*.

König, E.: *Augustinus Philosophus*, München, 1970.

Kuhn, Th. S.: *The Structure of Scientific Revolutions*, Chicago, 1970.

Lampey: *Das Zeitproblem nach den Bekenntnissen Augustinus*, Regensburg, 1960.

Lasserre, F. (ed.): *Die Fragmente des Eudoxos von Knidos*, Berlin, 1966.

Lechner, O.: *Idee und Zeit in der Metaphysik Augustinus*, München, 1964.

Lesky, A.: *Geschichte der griechischen Literatur*, Bern and München, 1962.
Lorenz, K.: 'Dialogspiele als semantische Grundlage von Logikkalkülen', in: *Archiv f. mathemat. Logik und Grundlagenforschung* 11/1–2 und 3–4, 1968.
Lorenz, K. and Mittelstraß, J. (eds.): *H. Dingler, Die Ergreifung des Wirklichen*, Kap. I–IV, Reihe Suhrkamp Theorie I, Frankfurt 1969.
Lorenzen, P.: *Einführung in die operative Logik und Mathematik*, Berlin-Göttingen-Heidelberg, 1955.
Lorenzen, P.: *Entstehung der exakten Wissenschaften*, Berlin-Göttingen-Heidelberg, 1960.
Lorenzen, P.: (ed.): *Hugo Dingler, Aufbau der exakten Fundamentalwissenschaft*, München, 1964.
Lorenzen, P.: *Differential und Integral*, Frankfurt 1965.
Lorenzen, P.: *Formale Logik*, Berlin 1967.
Lorenzen, P.: *Normative Logic and Ethics*, Mannheim, 1969.
Lorenzen, P.: *Methodisches Denken*, Frankfurt, 1974.
Lorenzen, P. and Schwemmer, O.: *Konstruktive Logik, Ethik and Wissenschaftstheorie*, Mannheim/Wien/Zürich 1973.
Lorenzen, P.: Wissenschaftstheorie und Wissenschaftssysteme, Vortrag zum 2. Internationalen Hegel-Kongreß, Stuttgart 1975.
Lorenzen, P.: *Hypotheses non fingo*, 1975.
Lorenzen, P.: *Elementargeometrie. Das Fundament der Analytischen Geometrie*, Mannheim/Wien/Zürich, 1984.
Lübke, A.: *Die Uhr*, Düsseldorf, 1958.
Luce, R. D. *et al.* (eds.): *Handbook of Mathematical Psychology*, New York, 1963.
Maier, A., 'Die Anfänge des physikalischen Denkens im 14. Jahrhundert', in: *Philosophia Naturalis* 1, 1950, 8–35.
Maier, A.: *Metaphysische Hintergründe der spätscholastischen Naturphilosophie*. Rome, 1955.
Maurice, K.: *Von Uhren und Automaten*, München, 1968.
Mittelstaedt, P.: *Philosophische Probleme der modernen Physik*, Mannheim, 1972.
Mittelstaedt, P.: *Klassische Mechanik*, Mannheim, 1970.
Mittelstaedt, P.: *Die Sprache der Physik*, Mannheim, 1972.
Mittelstaedt, P.: *Die Zeitbegriff in der Physik*, Zürich, 1976.
Mittelstaedt, P.: 'Ptotophysik und spezielle Relativitätstheorie', in: K. Lorenz (ed.), *Konstruktionen versus Positionen*, Bd. I, Berlin, New York, 1979, pp. 290–311.
Mittelstraß, J.: Die Entdeckung der Möglichkeit von Wissenschaft; in: *Archive for History of Exact Sciences* II, 1965, 410–435.
Mittelstraß, J.: *Neuzeit und Aufklärung. Studien zur Entstehung der neuzeitlichen Wissenschaft und Philosophie*, Berlin, New York, 1970.
Mittelstraß, J.: 'Wider den Dingler-Komplex', in: *Die Möglichkeit von Wissenschaft*, Frankfurt 1975, pp. 84–105.
Mittelstraß, J.: 'Die Prädikation und die Wiederkehr des Gleichen', in: *Ratio* **10**, 1966, 53–61 (engl. 78–87).
Moreau, J.: 'Le temps selon Aristote', *Revue philosophique de Louvain* **46**, 1948.
Müller-Markus, S.: *Protophysik, Entwurf einer Philosophie des Schöpferischen*, Den Haag, 1971.

Nagel, E.: *The Structure of Science*, New York, 1961.
Needham, J., Wang Ling, and de Solla Price, D. J.: *Heavenly Clockwork*, Cambridge, 1960.
Negro, W. v. del: 'Diskussionsbemerkung zum aristotelisch-augustinischen Zeitparadoxon', in: *Zeitschrift für Philosophische Forschung* **20**, 1966, 309–312.
Neurath, O.: 'Protokollsätze', in: *Erkenntnis* **3**, 1932.
Oetjens, H.: *Sprache, Logik, Wirklichkeit*, Stuttgart-Bad Cannstatt, 1975.
Pfarr, J.: 'Die Protophysik der Zeit und das Relativitätsprinzip', in: *Zeitschrift für allgemeine Wissenschaftstheorie* **VII**, 1976, 298–326.
Pfarr, J. (ed.): Protophysik und Relativitätstheorie. Mannheim/Wien/Zurich, 1981.
Pfarr, J.: 'Protophysics of Time and the Principle of Relativity', in: R. Cohen and M. Wartofsky, 1984, pp. 159–190.
Popper, K. R.: *Logik der Forschung*, Tübingen 1971.
Quinn, J. M.: 'The Concept of Time in St. Augustine', in: Studies in Philosophy and the History of Philosophy **4** (1969), pp. 75–127.
Reichenbach, H.: *Philosophie der Raum-Zeit-Lehre*, Berlin, 1928.
Robertson, J. Drummond: *The Evolution of Clockwork*, London, 1931.
Rochot, B.: 'Sur les notions de temps et d'espace chez quelques auterurs du XVII siècle, notamment Gassendhi et Barrow', in: *Revue d'histoire des sciences et de leurs applications*, 1956/57.
Sarton, G.: *A History of Science*. Cambridge, Massachusetts, 1959.
Schäfer, L.: *Erfahrung und Konvention. Zum Theoriebegriff der empirischen Wissenschaften*, Stuttgart-Bad Cannstatt, 1974.
Schmitt, B.: *Der Geist als Grund der Zeit. Die Zeitauslegung des Aurelius Augustinus*, Diss. Freiburg/Brsg., 1967.
Schneider, H. J.: *Historische und systematische Untersuchungen zur Abstraktion*, Diss. Erlangen, 1970.
Sneed, J. D.: *The Logical Structure of Mathematical Physics*, D. Reidel, Dordrecht, 1971.
Stegmüller, W.: *Hauptströmungen der Gegenwartsphilosophie*, Stuttgart, 1965.
Stegmüller, W.: *Probleme und Resultate der Wissenschaftstheorie und analytischen Philosophie*, II, 1, Berlin/Heidelberg/New York, 1970, II, 2, 1973.
Ströker, E.: *Einführung in die Wissenschaftstheorie*, München, 1973.
Suppes, P. and Zinnes, J.: 'Basic Measurement Theory', in R. D. Luce *et al.* (eds.), *Handbook of Mathematical Psychology*, New York, 1963.
Suter, R.: 'Augustinus on Time with Some Criticism from Wittgenstein', in: *Revue internationale de philosophie* **16** (1962), 387–394.
Taton, R. (ed.): *Ancient and Medieval Science*, New York, 1963.
Tetens, H.: 'Physik am normativen Gängelband? Über das Verhältnis der Protophysik zur empirischen Physik', *Zeitschrift für allgemeine Wissenschaftstheorie* **IV/1**, 1984, 142–160.
Thiel, Ch. (ed.): *Frege und die moderne Grundlagenforschung*, Meisenheim, 1975.
Thorndike, L.: 'Invention of the Mechanical Clock About 1271 A.D.', *Speculum* **16**, 1941.
Torstrik, A.: 'Über die Abhandlung des Aristoteles von der Zeit Phys. IV', in: *Philologus* **26**, 1867.
Tuomela, R.: *Theoretical Concepts*, Wien/New York, 1973.

Wagner, H.: *Aristoteles*, Berlin, 1967.
Weizsäcker, C. F. v.: 'Möglichkeit und Bewegung', in: *Festschrift für Joseph Klein*, Göttingen, 1967.
Weizsäcker, C. F. v.: 'Geometrie und Physik', in: Ch. P. Enz and J. Mehra (eds.), *Physical Reality and Mathematical Description*, D. Reidel, Dordrecht, 1974, pp. 48–90.
Weizsäcker, C. F. v.: *Die Einheit der Natur*, München, 1971.
Weyermann, A.: *Neue historisch-biographisch-artistische Nachrichten von Gelehrten und Künstlern*, Ulm, 1829.
Weyl, H.: *Raum, Zeit, Materie*, Berlin, 1923.
White, L.: '»Was beschleunigte den technischen Fortschritt im westlichen Mittelalter?'« In: *Technikgeschichte*, p. 32, 1965.
Wieland, W.: *Die Physik des Aristoteles*, Göttingen, 1962.
Whitrow, G. J.: *The Natural Philosophy of Time*, London/Edinburgh, 1961.
Wunderle, G.: 'Die Lehre des Aristoteles von der Zeit'. In: *Philosophisches Jahrbuch*, Bd. **XXI**, Fulda, 1908.
Zinner, E.: *Aus der Frühzeit der Räderuhr*, München, 1954.

NAME INDEX

SUBJECT INDEX

BOSTON STUDIES IN THE PHILOSOPHY OF SCIENCE

1. Marx W. Wartofsky (ed.), *Proceedings of the Boston Colloquium for the Philosophy of Science 1961-1962.* 1963.
2. Robert S. Cohen and Marx W. Wartofsky (eds.), *In Honor of Philipp Frank.* 1965.
3. Robert S. Cohen and Marx W. Wartofsky (eds.), *Proceedings of the Boston Colloquium for the Philosophy of Science 1964-1966. In Memory of Norwood Russell Hanson.* 1967.
4. Robert S. Cohen and Marx W. Wartofsky (eds.), *Proceedings of the Boston Colloquium for the Philosophy of Science 1966-1968.* 1969.
5. Robert S. Cohen and Marx W. Wartofsky (eds.), *Proceedings of the Boston Colloquium for the Philosophy of Science 1966-1968.* 1969.
6. Robert S. Cohen and Raymond J. Seeger (eds.), *Ernst Mach: Physicist and Philosopher.* 1970.
7. Milic Capek, *Bergson and Modern Physics.* 1971.
8. Roger C. Buck and Robert S. Cohen (eds.), *PSA 1970. In Memory of Rudolf Carnap.* 1971.
9. A. A. Zinov'ev, *Foundations of the Logical Theory of Scientific Knowledge (Complex Logic).* (Revised and enlarged English edition with an appendix by G. A. Smirnov, E. A. Sidorenka, A. M. Fedina, and L. A. Bobrova.) 1973.
10. Ladislav Tondl, *Scientific Procedures.* 1973.
11. R. J. Seeger and Robert S. Cohen (eds.), *Philosophical Foundations of Science.* 1974.
12. Adolf Grünbaum, *Philosophical Problems of Space and Time.* (Second, enlarged edition.) 1973.
13. Robert S. Cohen and Marx W. Wartofsky (eds.), *Logical and Epistemological Studies in Contemporary Physics.* 1973.
14. Robert S. Cohen and Marx W. Wartofsky (eds.), *Methodological and Historical Essays in the Natural and Social Sciences. Proceedings of the Boston Colloquium for the Philosophy of Science 1969-1972.* 1974.
15. Robert S. Cohen, J. J. Stachel and Marx W. Wartofsky (eds.), *For Dirk Struik. Scientific, Historical and Political Essays in Honor of Dirk Struik.* 1974.
16. Norman Geschwind, *Selected Papers on Language and the Brain.* 1974.
17. B. G. Kuznetsov, *Reason and Being: Studies in Classical Rationalism and Non-Classical Science.* (forthcoming).
18. Peter Mittelstaedt, *Philosophical Problems of Modern Physics.* 1976.
19. Henry Mehlberg, *Time, Causality, and the Quantum Theory* (2 vols.). 1980.
20. Kenneth F. Schaffner and Robert S. Cohen (eds.), *Proceedings of the 1972 Biennial Meeting, Philosophy of Science Association.* 1974.
21. R. S. Cohen and J. J. Stachel (eds.), *Selected Papers of Léon Rosenfeld.* 1978.
22. Milic Capek (ed.), *The Concepts of Space and Time. Their Structure and Their Development.* 1976.

23. Marjorie Grene, *The Understanding of Nature. Essays in the Philosophy of Biology.* 1974.
24. Don Ihde, *Technics and Praxis. A Philosophy of Technology.* 1978.
25. Jaakko Hintikka and Unto Remes, *The Method of Analysis. Its Geometrical Origin and Its General Significance.* 1974.
26. John Emery Murdoch and Edith Dudley Sylla, *The Cultural Context of Medieval Learning.* 1975.
27. Marjorie Grene and Everett Mendelsohn (eds.), *Topics in the Philosophy of Biology.* 1976.
28. Joseph Agassi, *Science in Flux.* 1975.
29. Jerzy J. Wiatr (ed.), *Polish Essays in the Methodology of the Social Sciences.* 1979.
30. Peter Janich, *Protophysics of Time.* 1985.
31. Robert S. Cohen and Marx W. Wartofsky (eds.), *Language, Logic, and Method.* 1983.
32. R. S. Cohen, C. A. Hooker, A. C. Michalos, and J. W. van Evra (eds.), *PSA 1974: Proceedings of the 1974 Biennial Meeting of the Philosophy of Science Association.* 1976.
33. Gerald Holton and William Blanpied (eds.), *Science and Its Public: The Changing Relationship.* 1976.
34. Mirko D. Grmek (ed.), *On Scientific Discovery.* 1980.
35. Stefan Amsterdamski, *Between Experience and Metaphysics. Philosophical Problems of the Evolution of Science.* 1975.
36. Mihailo Marković and Gajo Petrović (eds.), *Praxis. Yugoslav Essays in the Philosophy and Methodology of the Social Sciences.* 1979.
37. Hermann von Helmholtz, *Epistemological Writings. The Paul Hertz/Moritz Schlick Centenary Edition of 1921 with Notes and Commentary by the Editors.* (Newly translated by Malcolm F. Lowe. Edited, with an Introduction and Bibliography, by Robert S. Cohen and Yehuda Elkana.) 1977.
38. R. M. Martin, *Pragmatics, Truth, and Language.* 1979.
39. R. S. Cohen, P. K. Feyerabend, and M. W. Wartofsky (eds.), *Essays in Memory of Imre Lakatos.* 1976.
42. Humberto R. Maturana and Francisco J. Varela, *Autopoiesis and Cognition. The Realization of the Living.* 1980.
43. A. Kasher (ed.), *Language in Focus: Foundations, Methods and Systems. Essays Dedicated to Yehoshua Bar-Hillel.* 1976.
44. Trân Duc Thao, *Investigations into the Origin of Language and Consciousness.* (Translated by Daniel J. Herman and Robert L. Armstrong; edited by Carolyn R. Fawcett and Robert S. Cohen.) 1984.
46. Peter L. Kapitza, *Experiment, Theory, Practice.* 1980.
47. Maria L. Dalla Chiara (ed.), *Italian Studies in the Philosophy of Science.* 1980.
48. Marx W. Wartofsky, *Models: Representation and the Scientific Understanding.* 1979.
49. Trân Duc Thao, *Phenomenology and Dialectical Materialism.* 1985.
50. Yehuda Fried and Joseph Agassi, *Paranoia: A Study in Diagnosis.* 1976.
51. Kurt H. Wolff, *Surrender and Catch: Experience and Inquiry Today.* 1976.
52. Karel Kosík, *Dialectics of the Concrete.* 1976.
53. Nelson Goodman, *The Structure of Appearance.* (Third edition.) 1977.

54. Herbert A. Simon, *Models of Discovery and Other Topics in the Methods of Science.* 1977.
55. Morris Lazerowitz, *The Language of Philosophy. Freud and Wittgenstein.* 1977.
56. Thomas Nickles (ed.), *Scientific Discovery, Logic, and Rationality.* 1980.
57. Joseph Margolis, *Persons and Minds. The Prospects of Nonreductive Materialism.* 1977.
59. Gerard Radnitzky and Gunnar Andersson (eds.), *The Structure and Development of Science.* 1979.
60. Thomas Nickles (ed.), *Scientific Discovery: Case Studies.* 1980.
61. Maurice A. Finocchiaro, *Galileo and the Art of Reasoning.* 1980.
62. William A. Wallace, *Prelude to Galileo.* 1981.
63. Friedrich Rapp, *Analytical Philosophy of Technology.* 1981.
64. Robert S. Cohen and Marx W. Wartofsky (eds.), *Hegel and the Sciences.* 1984.
65. Joseph Agassi, *Science and Society.* 1981.
66. Ladislav Tondl, *Problems of Semantics.* 1981.
67. Joseph Agassi and Robert S. Cohen (eds.), *Scientific Philosophy Today.* 1982.
68. Władysław Krajewski (ed.), *Polish Essays in the Philosophy of the Natural Sciences.* 1982.
69. James H. Fetzer, *Scientific Knowledge.* 1981.
70. Stephen Grossberg, *Studies of Mind and Brain.* 1982.
71. Robert S. Cohen and Marx W. Wartofsky (eds.), *Epistemology, Methodology, and the Social Sciences.* 1983.
72. Karel Berka, *Measurement.* 1983.
73. G. L. Pandit, *The Structure and Growth of Scientific Knowledge.* 1983.
74. A. A. Zinov'ev, *Logical Physics.* 1983.
75. Gilles-Gaston Granger, *Formal Thought and the Sciences of Man.* 1983.
76. R. S. Cohen and L. Laudan (eds.), *Physics, Philosophy and Psychoanalysis.* 1983.
77. G. Böhme et al., *Finalization in Science*, ed. by W. Schäfer. 1983.
78. D. Shapere, *Reason and the Search for Knowledge.* 1983.
79. G. Andersson, *Rationality in Science and Politics.* 1984.
80. P. T. Durbin and F. Rapp, *Philosophy and Technology.* 1984.
81. M. Marković, *Dialectical Theory of Meaning.* 1984.
82. R. S. Cohen and M. W. Wartofsky, *Physical Sciences and History of Physics.* 1984.
83. E. Meyerson, *The Relativistic Deduction.* 1985.
84. R. S. Cohen and M. W. Wartofsky, *Methodology, Metaphysics and the History of Sciences.* 1984.
85. György Tamás, *The Logic of Categories.* 1985.
86. Sergio L. de C. Fernandes, *Foundations of Objective Knowledge.* 1985.
87. Robert S. Cohen and Thomas Schnelle (eds.), *Cognition and Fact.* 1985.
88. Gideon Freudenthal, *Atom and Individual in the Age of Newton.* 1985.

9 789027 707246

GPSR Compliance

The European Union's (EU) General Product Safety Regulation (GPSR) is a set of rules that requires consumer products to be safe and our obligations to ensure this.

If you have any concerns about our products, you can contact us on ProductSafety@springernature.com

In case Publisher is established outside the EU, the EU authorized representative is:

Springer Nature Customer Service Center GmbH
Europaplatz 3
69115 Heidelberg, Germany

Batch number: 10761599

Printed by Printforce, the Netherlands